Forest Soils and Carbon Storage

Forest Soils and Carbon Storage

Pawan Pattanaik

RANDOM PUBLICATIONS
NEW DELHI (INDIA)

Forest Soils and Carbon Storage

ISBN 978-93-5111-609-7

Published in 2015 in India by

RANDOM PUBLICATIONS

4376-A/4B, Gali Murari Lal, Ansari Road
New Delhi-110 002
Phone : +9111-43580356, 011-23289044, 011-43142548
e-mail: sales@randompublications.com,
info@randompublications.com, randomexports@gmail.com

Reprinted 2022

Type Setting by : Friends Media, Delhi-110089
Digitally Printed at : Replika Press Pvt. Ltd.

Preface

Glaciers have advanced and retreated across Massachusetts many times over the past several million years. The last glacier retreated over 10,000 years ago. Our soils are the product of the interaction of vegetation and climate on what the glaciers left behind. Most upland, forested soils have developed out of what is called `glacial till,` that is, soils that were created and mixed by glaciers moving across the landscape. They are typically stoney, fine-textured soils on top of bedrock. They may or may not have hardpan layers. We also have pockets of `glacial deposit` soils which were left by streams and rivers created by melting glaciers. These deposit soils are made of mixtures of sand and gravel and are found where there were edges of tongues of the retreating glaciers--mostly in river and stream valleys. Many hillsides have scattered pockets and bands of these soils among the predominantly glacial till soils. There are also lacustrine and alluvial soils. These are both silty soils, the former from deposits on old lake bottoms and the latter from flood deposits along rivers and streams. Most of these soils are in agricultural use.

Forest soils hold about one-third of the carbon stored in Earth's terrestrial ecosystems, but we still have much to learn about how management affects carbon accumulation and loss in forest soils. Since maintaining soil carbon storage is important for mitigating climate change, sustaining forest productivity, and protecting water quality, it is vital to understand how practices like forest fertilization, timber harvesting, and prescribed burns affect forest soil carbon storage. These practices are valuable tools in the acquisition and protection of the natural resources that forests provide, and the ever-increasing human need for forest resources demands a sound scientific basis to management.

I would like to thank my team for standing beside me throughout my career and writing this book. My special thanks go to "Random Publications" who have published the book.

– Pawan Pattanaik

Contents

1

Understanding the Forestry

Although most people know what a forest is, a definition of it which suits all cases is by no means easy to give. Manwood, in his treatise of the *Lawes of the Forest*, defines a forest as "a certain territory of woody grounds, fruitful pastures, privileged for wild beasts and fowls of forest, chase and warren, to rest and abide in, in the safe protection of the king, for his princely delight and pleasure."

This primitive definition has, in modern times, when the economic aspect of forests came more into the foreground, given place to others, so that forest may, in a general way, now be described as " an area which is for the most part set aside for the production of timber and other forest produce, or which is expected to exercise certain climatic effects, or to protect the locality against injurious influences." As far as conclusions can now be drawn, it is probable that the greater part of the dry land of the earth was, at some time, covered with forest, which consisted of a variety of trees and shrubs grouped according to climate, soil and configuration of the several localities. The conditions for an uninterrupted regeneration of the forest were favourable and the result was vigorous production by the creative powers of soil and climate.

Then came man and by degrees interfered, until in most countries of the earth the area under forest has been considerably reduced. The first decided interference was probably due to the establishment of domestic animals; men burnt the forest to obtain pasture for their flocks. Subsequently similar measures on an ever-increasing scale were employed to prepare the land for agricultural purposes. More recently enormous areas of forests were destroyed by reckless cutting and subsequent firing in the extraction of timber for economic purposes.

It will readily be understood that the distribution and character of the now remaining forests must differ enormously. Large portions of the earth are still covered with dense masses of tall trees, while others contain low scrub or grass land, or are desert. As a general rule, natural forests consist of a number of different species intermixed; but in some cases certain species, called gregarious, have succeeded in obtaining the upper hand, thus forming more or less pure forests of one species only. The number of species differs very much. In many tropical forests hundreds of species may be found on a comparatively small area, in other cases the number is limited. Burma has several thousand species of trees and shrubs, Sind has only ten species of trees.

Central Europe has about forty species and the greater part of northern Russia, Sweden and Norway contains forests consisting of about half a dozen species. Elevation above the sea acts similarly to rising latitude, but the effect is much more rapidly produced. Generally speaking, it may be said that the Tropics and adjoining parts of the earth, wherever the climate is not modified by considerable elevation, contain broad-leaved species, palms, bamboos, andc. Here most of the best and hardest timbers are found, such as teak, mahogany and ebony.

The northern countries are rich in conifers. Taking a section from Central Africa to North Europe, it will be found that south and north of the equator there is a large belt of dense hardwood forest; then comes the Sahara, then the coast of the Mediterranean with forests of cork oak; then Italy with oak, olive, chestnut, gradually giving place to ash, sycamore, beech, birch and certain species of pine; in Switzerland and Germany silver fir and spruce gain ground. Silver fir disappears in central Germany and the countries around the Baltic contain forests consisting chiefly of Scotch pine, spruce and birch, to which, in Siberia, larch must be added, while the lower parts of the ground are stocked with hornbeam, willow, alder and poplar.

In North America the distribution is as follows: Tropical vegetation is found in south Florida, while in north Florida it changes into a subtropical vegetation consisting of evergreen broad-leaved species with pines on sandy soils. On going north in the Atlantic region, the forest becomes temperate, containing deciduous broad-leaved trees and pines, until Canada is reached, where larches, spruces and firs occupy the ground. Around the great lakes on sandy soils the broad-leaved forest gives way to pines. On proceeding west from the Atlantic region the forest changes into a shrubby vegetation and this into the prairies.

Farther west, towards the Pacific coast, extensive forests are found consisting, according to latitude and elevation above the sea, of pines, larches, fir, Thujas and Tsugas. In Japan a tropical vegetation is found in the south, comprising palms, figs, ebony, mangrove and others. This is followed on proceeding north by subtropical forests containing evergreen oaks, *Podocarpus,* tree-ferns and, at higher elevations, *Cryptomeria* and *Chamaecyparis.* Then follow deciduous broad-leaved forests and finally firs, spruces and larches. In India the character of the forests is governed chiefly by rainfall and elevation.

Where the former is heavy evergreen forests of Guttiferae, Dipterocarpeae, Leguminosae, Euphorbias, figs, palms, ferns, bamboos and India-rubber trees are found. Under a less copious rainfall deciduous forests appear, containing teak and sal (*Shorea robusta*) and a great variety of other valuable trees. Under a still smaller rainfall the vegetation becomes sparse, containing acacias, *Dalbergia sissoo* and Tamarix.

Where the rainfall is very light or *nil,* desert appears. In the Himalayas, subtropical to arctic conditions are found, the forests containing, according to elevation, pines, firs, deodars, oaks, chestnuts, magnolias, laurels, rhododendrons and bamboos.

Australia, again, has its own particular flora of eucalypts, of which some two hundred species have been distinguished, as well as wattles. Some of the eucalypts attain an enormous height.

UTILITY OF FORESTS

In the economy of man and of nature forests are of direct and indirect value, the former chiefly through the produce which they yield and the latter through the influence which they exercise upon climate, the regulation of moisture, the stability of the soil, the healthiness and beauty of a country and allied subjects.

The *indirect* utility will be dealt with first. A piece of land bare of vegetation is, throughout the year, exposed to the full effect of sun and air currents and the climatic conditions which are produced by these agencies. If, on the other hand, a piece of land is covered with a growth of plants and especially with a dense crop of forest vegetation, it enjoys the benefit of certain agencies which modify the effect of sun and wind on the soil and the adjoining layers of air. These modifying agencies are as follows:

- The crowns of the trees intercept the rays of the sun and the falling rain; they obstruct the movement of air currents and reduce radiation at night.
- The leaves, flowers and fruits, augmented by certain plants which grow in the shade of the trees, form a layer of mould, or humus, which protects the soil against rapid changes of temperature and greatly influences the movement of water in it.
- The roots of the trees penetrate into the soil in all directions and bind it together. The effects of these agencies have been observed from ancient times and widely differing views have been taken of them.

Of late years, however, more careful observations have been made at so-called parallel stations, that is to say, one station in the middle of a forest and another outside at some distance from its edge, but otherwise exposed to the same general conditions. In this way, the following results have been obtained:

- Forests reduce the temperature of the air and soil to a moderate extent and render the climate more equable.
- They increase the relative humidity of the air and reduce evaporation.
- They tend to increase the precipitation of moisture. As regards the actual rainfall, their effect in low lands is *nil* or very small; in hilly countries it is probably greater, but definite results have not yet been obtained owing to the difficulty of separating the effect of forests from that of other factors.
- They help to regulate the water supply, produce a more sustained feeding of springs, tend to reduce violent floods and render the flow of water in rivers more continuous.
- They assist in preventing denudation, erosion, landslips, avalanches, the silting up of rivers" and low lands and the formation of sand dunes.
- They reduce the velocity of air-currents, protect adjoining fields against cold or dry winds and afford shelter to cattle, game and useful birds.
- They may, under certain conditions, improve the healthiness of a country and help in its defence.
- They increase the beauty of a country and produce a healthy aesthetic influence upon the people.

The *direct* utility of forests is chiefly due to their produce, the capital which they represent and the work which they provide. The principal produce of forests consists of timber and firewood. Both are necessaries for the daily life of the people. Apart from a limited number of broad-leaved species, the conifers have become the most important timber trees in the economy of man. They are found in greatest quantities in the countries around the Baltic and in North America. In modern times iron and other materials have, to a considerable extent, replaced timber, while coal, lignite and peat compete with firewood; nevertheless wood is still indispensable and likely to remain so. This is borne out by the statistics of the most civilized nations. Whereas the population of Great Britain and Ireland, during the period 1880-1900, increased by about 20 per cent, the imports of timber, during the same period, increased by 45 per cent; in other words, every head of population in 1900 used more timber than twenty years earlier. Germany produced in 1880 about as much timber as she required; in 1899 she imported 4,600,000 tons, valued at £14,000,000 and her imports are rapidly increasing, although the yield capacity of her own forests is much higher now than it was formerly. Wood is now used for many purposes which formerly were not thought of. The manufacture of the wood pulp annually imported into Britain consumes at least 2,000,000 tons of timber. A fabric closely resembling silk is now made of spruce wood. The variety of other, or minor, produce yielded by forests is very great and much of it is essential for the well-being of the people and for various industries.

The yield of fodder is of the utmost importance in countries subject to periodic droughts; in many places field crops could not be grown successfully without the leaf-mould and brushwood taken from the forests. As regards industries, attention need only be drawn to such articles as commercial fibre, tanning materials, dye-stuffs, lac, turpentine, resin, rubber, guttapercha, and c. Great Britain and Ireland alone import every year such materials to the value of £12,000,000, half of this being represented by rubber.

The *capital* employed in forests consists chiefly of the value of the soil and growing stock of timber. The latter is, ordinarily, of much greater value than the former wherever a sustained annual yield of timber is expected from a forest. In the case of a Scotch pine forest, for instance, the value of the growing stock is, under the above-mentioned condition, from three to five times that of the soil. The rate of interest yielded by capital invested in forests differs, of course, considerably according to circumstances, but on the whole it may, under proper management, be placed equal to that yielded by agricultural land; it is lower than the agricultural rate on the better classes of land, but higher on the inferior classes. Hence the latter are specially indicated for the forest industry and the former for the production of agricultural crops. Forests require *labour* in a great variety of ways, such as:

- General administration, formation, tending and harvesting;
- Transport of produce;
- Industries which depend on forests for their prime material. The labour indicated under the first head differs considerably according to circumstances, but its amount is smaller than that required if the land is used for agriculture. Hence forests provide additional labour only if they are established on surplus lands. Owing to the bulky nature of forest produce its transport forms a business of considerable magnitude, the amount of labour being perhaps equal to half that employed under the first head. The greatest amount of labour is, however, required in the working up of the raw material yielded by forests. In this respect attention may be drawn to the chair industry in and around High Wycombe in Buckinghamshire, where more than 20,000 workmen are employed in converting the beech, grown on the adjoining chalk hills, into chairs and tools of many patterns.
- Complete statistics for Great Britain are not available under this head, but it may be mentioned that in Germany the people

employed in the forests amount to 2.3 per cent of the total population; those employed on transport of forest produce. 1 per cent; labourers employed on the various wood industries, 8.6 per cent; or a total of 12 per cent. An important feature of the work connected with forests and their produce is that a great part of it can be made to fit in with the requirements of agriculture; that is to say, it can be done at seasons when field crops do not require attention. Thus the rural labourers or small farmers can earn some money at times when they have nothing else to do and when they would probably sit idle if no forest work were obtainable.

Whether, or how far, the utility of forests is brought out in a particular country depends on its special conditions, such as:

- The position of a country, its communications and the control which it exercises over other countries, such as colonies;
- The quantity and quality of substitutes for forest produce available in the country;
- The value of land and labour and the returns which land yields if used for other purposes;
- The density of population;
- The amount of capital available for investment;
- The climate and configuration, especially the geographical position, whether inland or on the border of the sea, and c. No general rule can be laid down, showing whether forests are required in a country, or, if so, to what extent; that question must be answered according to the special circumstances of each case.

These data exhibit considerable differences, since the percentage'of the forest area varies from 3.5 to 50 and the area per head of population from 07 to 9.5 acres. Russia, Sweden and Norway may as yet have more forest than they require for their own population.

On the other hand, Great Britain and Ireland, Germany, Denmark, Portugal, Holland and even Belgium, France and Italy have not a sufficient forest area to meet their own requirements; at the same time, they are all sea-bound countries and importation is easy, while most of them are under the influence of moist sea winds, which reduces to a subordinate position the importance of forests for climatic reasons.

Intimately connected with the area of forests in a country is the state of ownership - whether they belong to the state, corporations or to private persons. Where, apart from the financial aspect and the supply of work, forests are not required for the sake of their indirect effects and where importation from other countries is easy and assured, the government of the country need not, as a rule, trouble itself to maintain or acquire forests. Where the reverse conditions exist and especially where the cost of transport over long distances becomes prohibitive, a wise administration will take measures to assure the maintenance of a suitable proportion of the country under forest. This can be done either by maintaining or constituting a suitable area of state forests, or by exercising a certain amount of control over corporation and even private forests. Such measures are more called for in continental countries than in those which are sea-bound, as is proved by the above statistics.

SUPPLY OF TIMBER

Imports and Exports

The following table shows the net imports and exports of European countries (average data, calculated from the returns of recent years). The only timber-exporting countries of Europe are Russia, Sweden, Norway, Austria-Hungary and Rumania; all the others either have only enough for their own consumption, or import timber. Great Britain and Ireland import now upwards of 10,000,000 tons a year, Germany about 4,600,000 tons and Belgium about 1,300,000 tons. Holland, France, Portugal, Spain and Italy are all importing countries, as also are Asia Minor, Egypt and Algeria. The west coast of Africa exports hardwoods and imports coniferous timber. The Cape and Natal import considerable quantities of pine and fir wood. Australasia Net Imports and Exports of European Countries. These net imports are received from non-European countries. They consist chiefly of valuable hardwoods, like teak, mahogany, cucalypts and others.

Exports hardwoods and some Kauri pine from New Zealand, but imports larger quantities of light pine and fir timber. British India and Siam export teak and small quantities of fancy woods. The West Indies and South America export hardwoods and import pine and fir wood. The United States of America will not much longer be a genuine exporting country, since they import already almost as much timber from Canada as they export. Canada exports considerable quantities of timber. The Dominion has still a forest area of 1,250,000 sq. m., equal to 38 per cent of the total area and giving 165 acres of forest for every inhabitant. Although only about one-third of the forest area can be called regular timber land, Canada possesses an enormous forest wealth, with which she might supply permanently nearly all other countries deficient in material, if the governing bodies in the several provinces would only determine to stop the present fearful waste caused by axe and fire and to introduce a regular system of management. As matters stand, the supplies of the most valuable timber of Canada, the white or Weymouth pine, are nearly exhausted, the great stores of spruce in the eastern provinces are being rapidly destroyed and the forests of Douglas fir in the western provinces have been attacked for export to the United States and to other countries. Taking the remaining stocks of the whole earth together, it may be said that a sufficient quantity of hardwoods is available, but the only countries which are able to supply coniferous timber for export on a considerable scale are Russia, Sweden, Norway, Austria and Canada. As these countries have practically to supply the rest of the world and as the management of their forests is far from satisfactory, the question of supplying light pine and fir timber, which forms the very staff of life of the wood industries, must become a very serious matter before many years have passed.

Unmistakable signs of the coming crisis are everywhere visible to all who wish to see and it is difficult to over-state the gravity of the problem, when it is remembered, for instance, that 87 per cent of all the timber imported into Great Britain consists of light pine and fir and that most of the other importing countries are similarly situated.

In some of these countries little or no room exists for the extension of woodland, but this statement does not apply to Great Britain and Ireland, which contain upwards of 12,000,000 acres of waste land and 12, 500,000 acres of mountain and heath land used for light grazing. Onefourth of that area, if put under forest, would produce all the timber now imported which can be grown in Britain, that is to say, about 95 per cent of the total.

Forest Management

In early times there was practically no forest management. As long as the forests occupied considerable areas, their produce was looked upon as the free gift of nature, like air and water; men took it, used it and even destroyed it without let or hindrance. With the gradual increase of population and the consequent reduction of the forest area, proprietary ideas developed; people claimed the ownership of certain forests and proceeded to protect them against outsiders. Subsequently the law of the country was called in to help in protection, leading to the promulgation of special forest laws. By degrees it was found that mere protection was not sufficient and that steps must be taken to enforce a more judicious treatment, as well as to limit the removal of timber to what the forests were capable of producing permanently.

The teaching of natural science and of political economy was brought to bear upon the subject, so that now forestry has become a special science. This is recognized in many countries, amongst which Germany stands first, closely followed by France, Austria, Denmark and Belgium. Of non-European countries the palm belongs to British India and then follow Ceylon, the Malay States, the Cape of Good Hope and Japan.

The United States of America have also turned their attention to the subject. Most of the British colonies are, in this respect, as yet in a backward state and the matter has still to be fought out in Great Britain and Ireland, though many writers have urged the importance of the question upon the public and the government. There can be no doubt that all civilized countries must, sooner or later, adopt a rational and systematic treatment of their forests.

For details as to the separate countries, see the articles under the country headings; in this article only some of the more important countries are dealt with, in so far as the history of their forestry is important. A few notes on Germany and France will be given, because in these countries forest management has been brought to highest perfection; Italy is mentioned, because she has allowed her forests to be destroyed; and a short description of forestry in the United Kingdom and in India follows.

The winters being long and severe, an abundant supply of fuel is almost as essential as a sufficient supply of food. This necessity has led, along with a passion for the chase, to the preservation of forests and to the establishment of an admirable system of forest cultivation, almost as carefully conducted as field tillage. The Black Forest stretches the whole length of the grand-duchy of Baden and part of the kingdom of Wurttemberg, from the Neckar to Basel and the Lake of Constance. The vegetation resembles that of the Vosges; forests of spruce, silver fir, Scotch pine and, mingled with birches, beech and oak, are the chief woods met with. Until comparatively recent times large quantities of timber derived from these forests were floated down the Rhine to Holland and also shipped to England.

Now the greater part of it is used locally for construction, or it is converted into paper pulp. In the grand-duchy of Hesse the Odenwald range of mountains, stretching between the Main and the Neckar, contains the chief supply of timber. In the province of Nassau there are the large wooded tracts of the Taunus mountain range and the Westerwald.

In Rhenish Prussia valuable forests lie partly in the Eifel, on the borders of Belgium and on the mountains overhanging the Upper Moselle, but they do not furnish such stately trees as the Black Forest and the Odenwald. The Spessart, near Aschaffenburg in Bavaria, is one of the most extensive forests of middle Germany, containing large masses of fine oak and beech, with plantations of coniferous trees, such as spruce, Scotch pine and silver fir.

Bavaria possesses other fine forest tracts, such as the Baierischewald on the Bohemian frontier, the Kranzberg near Munich and the Frankenwald in the north of the kingdom. North Germany has extensive forests on the Harz and Thuringian Mountains, while in East Prussia large tracts of flat ground are covered with Scotch pine, spruce, oak and beech.

Every German state has its forest organization: In Prussia the department is presided over by the Oberland Forstmeister at Berlin, while each province, or part of a province, has an Oberforstmeister, under whom a number of

OberfOrsters administrate the state and communal forests. These, again, are assisted by a lower class of officials called Forsters. The Oberf6rsters throughout Germany are educated at special schools of forestry, of which in 1909 the following nine existed: In Prussia: at Eberswalde and Munden.

THE NATIONAL FOREST POLICY

The forest policy of the United States may be said to have had its origin in 1799 in the enactment of a law which authorized the purchase of timber suitable for the use of the navy, or of land upon which such timber was growing. It is true that laws were in force under the early governments of Massachusetts, New Jersey and other colonies, providing for the care and protection of forest interests in various ways, but these laws were distinctly survivals of tendencies acquired in Europe and for the most part of little use.

It was not until the apparent approach of a dangerous shortage in certain timber supplies that the first real step in forest policy was taken by the United States. Successive laws passed from 1817 to 1831 strove to give larger effect to the original enactment, but without permanent influence towards the preservation of the live oak, which was the object in view. A long period of inaction followed these early measures. *The unshaded areas are treeless, except along the Streams* 1831 the solicitor of the treasury assumed a partial responsibility for the care and protection of the public timber lands and in 18 55 this duty was transferred to the commissioner of the general land office in the Department of the Interior.

The effect of these changes upon forest protection was unimportant. When, however, at the close of the Civil War railway building in the United States took on an unparalleled activity, the destruction of forests by fire and the axe increased in a corresponding ratio and public sentiment began to take alarm. Action by several of the states slightly preceded that of the Federal government, but in 1876 Congress, acting under the inspiration of a memorial from the American Association for the Advancement of Science, authorized the appointment of an officer under the commissioner of agriculture, to collect and distribute information upon forest matters.

As the railways advanced into the treeless interior, public interest in tree-planting became keen. In 1873 Congress passed and later amended and repealed the timber culture acts, which granted homesteads on the treeless public lands to settlers who planted one-fourth of their entries with trees. Though these measures were not successful in themselves they directed attention towards forestry. The act which repealed them in 1891 contained a clause which lies at the foundation of the present forest policy of the United States.

By it the president was authorized to set aside " any part of the public lands wholly or in part covered with timber or undergrowth, whether of commercial value or not, as public reservations and the President shall, by public proclamation, declare the establishment of such reservations and the limits thereof." Some eighteen million :acres had been proclaimed as reservations at the time when, in 1896, the National Academy of Sciences was asked by the secretary of the interior to make an investigation and report upon " the inauguration of a rational forest policy for the forest lands of the United States." Upon the recommendation of a commission named by the Academy, President Cleveland established more than twenty-one million acres of new reserves on the 22nd of February 1897. His action was widely misunderstood and attacked, but it awakened a public interest in forest questions without which the rapid progress of forestry in the United States since that time could never have been made.

Within a few months after the proclamation of the Cleveland reserves the present national forest policy took definite shape. Under this policy the national government holds and manages, in the common interest of all users of the forests or its products, such portions of the public lands as have been set aside by presidential proclamation in accordance with the act of 1891.

These lands are held against private acquisition under the Homestead Act (except as to agricultural lands as hereafter mentioned), the Timber and Stone Act and other laws under which the United States disposes of its unappropriated public domain, but not against private acquisition under the Mineral Land Laws. They are selected from lands believed to be more valuable for forest purposes than for agriculture and are managed with the purpose of securing from them the best and largest possible returns, present and future, whether in the form of water for irrigation or power, of timber, of forage for stock, or of any other beneficial product. The aggregate area of the reserves, or national forests, has been steadily increased until they now include nearly all the timber lands left of the public domain.

The general lines of this policy were in part laid down by the commission already mentioned, in its report submitted to the secretary of the interior, May 1, 1897 and by the act of June 4, 1897, which was largely shaped by the work of the commission. Until this act was passed the national forests had been in theory closed against any form of use; nor had the possibility of securing forest preservation by wise use received much thought from those who had favoured their creation. Such a state of affairs could not continue.

Before long public opinion would have forced the opening to use of the resources thus :arbitrarily locked up and in the absence of any administrative system providing for conservative use, the national forests would inevitably have been abolished and the whole policy of government forest holdings would have ceased.

The act of June 4, 1897 was therefore of the first importance. This act conferred upon the secretary of the interior general powers for the proper management of the national forests through the general land office of his department. It provided for the designation and sale of dead, mature and large timber; authorized the secretary to permit free use of timber in small quantities by settlers, miners and residents; empowered him to "make such rules and regulations and establish such service as will insure the objects of such reservations, namely, to regulate their occupancy and use and to preserve the forests thereon from destruction "; and made violation of the act or of such rules and regulations a misdemeanour. The statute limited the power to establish forest reservations to the purpose of improving and protecting the forest, securing favourable conditions of water flows and furnishing a continuous supply of timber for the use and necessities of citizens of the United States.

Lands found, upon due examination, to be more valuable for other purposes than for forest uses might be eliminated from any reservation and all mineral lands within the reservations were left open to private appropriation under the mineral laws. The rights of settlers and claimants were safeguarded and civil and criminal jurisdiction, except so far as the punishment of offences against the United States in the reservations was concerned, was reserved to the States.

While the administration of the national forests was entrusted to the general land office, the same act assigned the surveying and mapping of them to the United States Geological Survey, which has published descriptions and maps of some of the more important.

No attempt was made in the general land office to develop a technical forest service. There were, indeed, at the time of passage of the act, less than ten trained foresters in the United States, no means of training more and very little conception of what forestry actually meant. The purpose of the administration was therefore mainly protection against trespass and fire, particularly the latter.

Regulations were made giving effect to the provisions of the act of June 4, set forth above, but in the absence of technical knowledge as to what might safely be done, the tendency was rather to restrict than to extend the use of the forest. Meanwhile, however, there was rapidly developing in another branch of the government service an organization qualified for actual forest management.

One year after the passage of the act of June 4, 1897, the division of forestry in the Department of Agriculture ceased to be merely a bureau of information and became an active agency for introducing the actual practice of forestry among private owners and for conducting the investigations upon which a sound American forest practice could be based. The work awakened great interest among forest owners and exerted a powerful educational influence upon the country at large. The division extended its work and became the Bureau of Forestry. It drew into its employment for a time nearly all the men who were preparing themselves in increasing numbers (at first abroad, then in the newly-founded schools in the United States) for the profession of forestry and was soon recognized as qualified to speak authoritatively on technical questions connected with the administration of the national forests.

This led to a request from the secretary of the interior for the advice of the bureau on such questions. Working plans were accordingly undertaken for a number of the forests. The general land office, however, was not ready to attempt active forest management. Though some timber was sold and the grazing of stock regulated to some extent, the main object of the land office administration continued to be protection against fire. Many of the regulations which it made could not be enforced.

The disadvantages of dispersal of the Federal government forest work among three separate agencies grew more and more apparent, until, on the 1st of February 1905, control of the 63,000,000 acres of forest reserves which up to that time had been set aside was transferred from the general land office to the Bureau of Forestry. In recognition of its new duties the designation of the bureau became the Forest Service.

Other provisions of the act which affected the transfer were that forest supervisors and rangers should be selected, so far as possible, from qualified citizens of the state or territory in which each forest was situated and that all money received from the sale of any products or the use of any land or resources of the national forests should be covered into the treasury and constitute a special fund for their protection, administration, improvement and extension.

Five days later a statute gave forest officers the power to arrest trespassers; and on the 3rd of March the lieu land selection law was repealed. This law had opened the way for grave abuses through the exchange of worthless land by private owners within the forests for an equal area of valuable timber lands outside.

The law has been modified since by the change of the old name " Forest Reserves " to " National Forests." The act of June iz, 1906, opened to homestead entry lands within national forests found by examination to be chiefly valuable for agriculture. The administration and improvement of the national forests are now provided for directly by congressional appropriation. The power to create national forests conferred on the president by the act of March 1891 has been repealed for the states of Washington, Oregon, Idaho, Montana, Wyoming and Colorado, but for no others.

The Forest Service began in earnest the development of all the resources of the national forests. Mature timber was sold wherever there was a demand for it and the permanent welfare of the forests and protection of the streams permitted, but always so as to prevent waste, guard against fire, protect young. growth and ensure reproduction. Regulations were adopted which allowed small sales to be made without formality or delay,. secured for the government the full value of timber sold and eliminated unnecessary routine.

Care was taken to safeguard the interests of the government and provide for the maintenance of good technical standards. The conduct of local business was entrusted to local officers. Large transactions with general policies were controlled from Washington, but with careful 2 = *District Boundaries and Numbers* provision for first-hand knowledge and close touch with the work in the field.

Business efficiency and the convenience of the public were carefully studied. In short, an organization was created capable of handling safely, speedily and satisfactorily the complex business of making useful a forest property of vast extent, scattered through sixteen different states of an aggregate area of over 1,50o,000 sq. m. and with a population of 9,000,000.

The growth since the 1st of July 1897 of the area of the national forests, of the expenditures of the government for forestry and of the receipts from the national forests, is shown by the statement which follows. Though the act of June 4, 1897, became effective immediately upon its passage, the fiscal year 1899 was the first of actual administration, because the first for which Congress made the appropriation necessary to carry out the law.

Forest is ripe for the axe, the demand is strong and control by trained men makes it safe to cut more freely. The increase is marked both in small and in large sales, but a score of sales for less than $5000 are made against one for more. The total cut is still far below the annual increment of the forests. As the demand grows restrictions must increase in order to husband the present supply until the next crop matures.

- The stumpage price would seem on the face of the figures to have risen from about one dollar to more than three dollars per thousand board-feet. The receipts, however, for any one year are not exclusively for the timber cut in that year, since payments are made in advance. In the year 1907 the average price obtained was something less than $2.50 per thousand. It is therefore true that stumpage prices have risen greatly, although conditions new to the American lumbermen are im *Area of National Forests, Annual Expenditures of the Federal Government for Forestry and National Forest Administration and Receipts from National Forests, 1898-1909* Until 1906, the sole source of receipts was the sale of timber. In the fiscal year 1907, however, timber sales furnished less than half the receipts. The following statement concerning the timber sales of the fiscal years 1904-1907 will serve to bring out the change that followed the transfer of control to the forest service in the midst of the fiscal year 1905:

  - A large excess each year in the amount of timber sold over that cut and paid for;
  - Nine times as much timber sold at the end of the four-year period as at the beginning and three times as much cut;
  - A much higher price obtained per thousand board-feet at the end of the period than at the beginning. Each of these matters calls for comment. The sales are of stumpage only; the government does no logging on its own account.
- More timber is sold each year than is cut and paid for, because many of the sales extend over several years. With increasing sales the amount sold each year for future removal has exceeded the amount to be removed during that year under sales of earlier years. Large sales covering a term of years are made because the national forests contain much overmature timber, which needs removal, but which is frequently too inaccessible to be saleable in small amounts. To prevent speculation the time allowed for cutting is never more than five years and cutting must begin at once and be continued steadily.
- The volume of sales has increased rapidly because much 1 The United States fiscal year ends June 30 and receives its designation from the calendar year in which it terminates. Thus, the fiscal year 1898 is the year July 1, 1897 - June 30, 1898.

Administration transferred to Bureau of Forestry, February I, 1905. Full utilization of all merchantable material, care of young growth in felling and logging and the piling of brush, to be subsequently burned by the forest officers if burning is necessary, are among these conditions.

Timber to be cut must first be marked by the forest officers. Sales of more than $100 in value are made only after public advertisement.

Only the simplest forms of silviculture have as yet been introduced. The vast area of the national forests, the comparatively sparse population of the West, the rough and broken character of the forests themselves and the newness of the problems which their management presents, make the general application of intensive methods for the present impracticable.

Natural reproduction is secured. The selection system is most used, often under the rough and ready method of an approximate diameter limit, with the reservation of seed trees where needed. The tendency, however, is strongly towards a more flexible and effective application of the selection principle, as a better trained field force is developed and as market conditions improve.

One conspicuous achievement was the reduction of loss by fires on the national forests. During the unusually dry season of 1905 there were only eight fires of any importance and the area burned over amounted only to about 16 of 1 per cent of the total area. In 1906 about 12 of 1 per cent was burned. This was accomplished by efficient patrol, co-operation of the public and by preventive measures, such as piling and burning the brush on cut-over areas.

Since the beginning of 1906 the largest source of income from the national forests was their use for grazing. Stock-raising is one of the most important industries of the West. Formerly cattle and sheep grazed freely on all parts of the public domain. In the early days of the national forests the wisdom of permitting any grazing at all upon them was sharply questioned.

Unrestricted grazing had led to friction between individuals, the deterioration of much of the range through overstocking and serious injury to the forests and stream flow. The forests of the West, however, are largely of open growth and contain many grassy parks, the results of old fires and many high mountain meadows.

Under proper regulations the grass and other forage plants which they produce in great quantity can be used without detriment to the forests themselves and with great benefit to the stock industry, which often can find summer pasturage nowhere else.

Except in southern California grazing is now permitted on all national forests unless the watersheds furnish water for domestic use; but the time of entering and leaving, the number of head to be grazed by each applicant and the part of the range to be occupied are carefully prescribed. Planted areas and cut-over areas are closed to stock until the young growth is safe from harm and goats are allowed only in the brushland of the foothills.

The results of regulation, in addition to the protection of forest growth and streams, are the prevention of disputes, improved range, better stock, stable conditions in the stock industry and the best use of the range in the interest of progress and development. The first right to graze stock on the forests is given to residents, small owners and those who have used the range before. Thus the crowding out of the weaker by the stronger and of the settler by the roving outsider has been stopped. In 1906 the forest service began to impose a moderate charge for the use of the national forest range.

The following statement shows the amount of stock grazed on the national forests 1904-09 and the receipts for the grazing charge: - A work of enormous magnitude which has now begun is planting on the national forests. At present, with low stumpage prices and incomplete utilization of forest products, clear cutting with subsequent planting is not practicable.

There are, however, many million acres of denuded land within the national forests which require planting. Such planting is still confined chiefly to watersheds which supply cities and towns with water. The first planting was done in 1892, in California. Since then similar work has been done on city watersheds in Colorado, Utah, Idaho and New Mexico. Other plantations are in the Black Hills national forest, where large areas of cut-over and burned-over land are entirely without seed trees and in the sandhill region of Nebraska. Up to 1908 about 2,000,000 seedlings had been planted, on over 2000 acres - a small beginning, but the work was entirely new and presented many hard problems.

The nursery operations of the forest service are concentrated at seven stations, located in southern California, Nebraska, Colorado, New Mexico Utah and Idaho, where stock is raised for local planting and for shipment elsewhere. These nurseries are small. Their annual productive capacity is between 8,000,000 and seedlings. Each nursery is practically an experimental forest-planting station, at which a large variety of species are grown and various methods are tried.

The organization of the administrative work of the national forests is by single forests. On the 1st of January 1908 the total number of forests was 165 with a total area of 162,023,190 acres. In charge of each forest is a forest supervisor. Under the supervisors are forest rangers and forest guards, whose duties include patrol, marking timber and scaling logs, enforcing the regulations and conducting some of the minor business arising from the use of the forests. Guards are temporary employes; rangers are employed by the year. The supervisors report directly to and receive instructions from the central office at Washington.

In this office there are four branches - operation, grazing, silviculture and products - each of which directs that part of the work which belongs to it, dealing directly with the supervisor.

For inspection purposes, however, the forests are separated into six districts, in each of which is located a chief inspector with a corps of assistants. The inspectors are without administrative authority, but assist by their counsel the supervisors and through inspection reports keep the Washington office informed of the condition of all lines of administrative work in progress. Administrative officers alternate frequently between field and office duties.

The number of forest officers in the several grades on the 1st of January 1908 were: 6 chief inspectors, 26 inspectors, 106 forest supervisors, 41 deputy forest supervisors, 820 forest rangers and 283 forest guards. The total number of employes of the forest service on the same date, including the clerical force, was 2034.

Besides the administration of the national forests, the forest service conducts general investigations, carries on an extensive educational work and co-operates with private owners who contemplate forest management upon their own tracts. This last work is undertaken because of the need of bringing forestry into practice, the lack of trained foresters outside of the employ of the government and the lack of information as to how to apply forestry and what returns may be obtained. Co-operation takes the form of advice upon the ground and, on occasion, of the making of working plans.

The educational work of the service is performed chiefly through publications, the purpose of which is to spread very widely a knowledge of the importance of forestry to the nation and of the principles upon which its practice rests. The investigations which the service conducts extend from studies. of the natural distribution and classification of American forests. and of their varied silvicultural problems to statistics of lumber production and laboratory researches which bear upon the economical utilization of forest products. As examples of these researches may be mentioned tests of the strength of timber, studies of the preservative treatment of wood for various uses, wood-pulp investigations and studies in wood chemistry.

Forestry Among the States

Among the states forestry has hardly reached the stage of practical application on the ground. New York holds 1,500,00o acres of forest land. It has a commission to care for its forest preserve and to protect the forest land throughout the state from fire. The constitution of the state, however, prohibits the cutting of timber on state land and thus confines the work entirely to protection of the forest and to the planting of waste areas. Pennsylvania is at present showing the most efficient activity in working out a forest policy. It has state forests of 820,000 acres, a good fire law more and more satisfactorily enforced and eight nurseries for growing planting material. In 1905, 160,000 white pine seedlings were set out. It has also a school for forest rangers, to be employed on the state forests and it has just established a state professional school of forestry.

Twenty-six of the states have regularly appointed forest officers, six have carried on studies of forest conditions in co-operation with the forest service and there is scarcely one which is not actively interested in forestry. Laws, generally good, to prevent damage from forest fires, have been enacted by practically all the states, but their enforcement has unfortunately been lax.

Public sentiment, however, is making rapid progress. Among the best laws are those of Maine, New Hampshire, Minnesota, New York, Pennsylvania and Wisconsin. The New York law, for example, provides for the appointment of one or more firewardens in each town of the counties in which damage by fire is especially to be feared. In other counties supervisors of towns are *ex-officio* fire-wardens. A chief fire-warden has general supervision of their work.

The wardens, half of the cost of whose services is paid by the state, receive compensation only for the time actually employed in fighting fires. They may command the service of any citizen to assist them. Setting fire to woods or waste lands belonging to the state or to another, if such fire results in loss, is punishable by a fine not exceeding $250 or imprisonment not exceeding one year, or both and damages are provided for the person injured. Since fire is beyond question the most dangerous enemy of forests in the United States, the measures taken against it are of vital importance.

The following table shows the amount of forest land held by the different states and by the territory of Hawaii: A *rea of State Forest Reservations, 1907. Forestry on Private Lands. The* practice of forestry among private owners is of old date. One of the earliest instances was that of Jared Eliot, who, in 1730, began the systematic cutting of timber land to supply charcoal for an iron furnace at Old Salisbury, Connecticut. The successful planting of waste lands with timber trees in Massachusetts dates from about ten years later. But such examples were comparatively rare until recent times.

At present the intelligent harvesting of timber with a view to successive crops, which is forestry, is much more common than is usually supposed. Among farmers it is especially frequent. It was begun among lumbermen by the late E. S. Coe, of Bangor, Maine, who made a practice of restricting the cut of spruce from his forests to trees ro, 12 or sometimes even 14 in. in diameter, with the result that much of his land yielded, during his life, a second crop as plentiful as the first.

Many owners of spruce lands have followed his example, but until very recently without improving upon it. Systematic forestry on a large scale among lumbermen was begun in the Adirondacks during the summer of 1898 on the lands of Dr W. S. Webb and Hon. W. C. Whitney, of a combined area of over ioo,000 acres, under the superintendence of the then Division of Forestry. In these forests spruce, maple, beech and birch predominate, but the spruce alone is at present of tht first commercial importance.

The treatment is a form of the selection system. Under it a second crop of equal yield would be ripe for the axe in thirtyfive years. Spruce and pine are the only trees cut. The work had been executed, at least up to the year 1902, with great satisfaction to the owners and the lumbering contractors, as well as to the decided benefit of the forest. The lumbering is regulated by the following rules and competent inspectors are employed to see that they are rightly carried out:

- No trees shall be cut which are not marked.
- All trees marked shall be cut.
- No trees shall be left lodged in the woods and none shall be overlooked by the skidders or haulers.
- All merchantable logs which are as large as 6 in. in diameter at the small end must be utilized.
- No stumps shall be cut more than 6 in. higher than the stump is wide.

- No spruce shall be used for bridges, corduroy, skids, slides, or for any purpose except building camps, dams or booms, unless it is absolutely necessary on account of lack of other timber.
- All merchantable spruce used for skidways must be cut into logs and hauled out.
- Contractors must not do any unnecessary damage to young growth in lumbering; and if any is done, they must discharge the men who did it.

These two instances of forestry have been most useful and effective among lumbermen and other owners of forest land in the north-east. Among those which have followed their example are the Berlin Mills Paper Company in northern New Hampshire,, the Cleveland Cliffs Iron Company in northern Michigan and the Delaware and Hudson Railroad Company in New York, all of which have employed professional foresters.

The most notable instance of forestry in the south is on the estate of George W. Vanderbilt at Biltmore, N.C. This was the first case of systematic forestry under regular working plans in the United States. It was begun in 1891 on about 4000 acres and has since been extended until it now covers about ioo,000 acres. A professional forester with a corps of trained rangers under him is in charge of the work. The Pennsylvania Railroad has recently employed a trained forester and several assistants and has undertaken systematic forestry on a large scale.

The effect of the work of the forest service in assisting private owners is evidenced by the fact that down to the year 1908 670 wood lots and timber tracts had been examined by agents of the forest service, of which 250 were tracts over 400 acres in extent and planting plans had been made for 436 owners covering a total area of 80,000 acres. Expert advice is also given to wood lot owners upon application by many of the state foresters.

American Practice

The conditions under which forestry is practised in Europe and in America differ so widely that rules which are received as axiomatic in the one must often be rejected in the other. Among these conditions in America are the highly developed and specialized methods and machinery of lumbering, the greater facilities for transportation and consequent greater mobility of the lumber trade, the vast number of small holdings of forest land and the enormous supply of low-grade wood in the timbered regions.

High taxes on forest properties, cut-over as well as virgin, notably in the north-western pineries and the firmly established habits of lumbermen, are factors of great importance. From these and other considerations it follows that such generally accepted essentials of European methods of forestry as a sustained annual yield, a permanent force of forest labourers, a permanent road system and the like, are in most cases utterly inapplicable in the United States at the present day in private forestry. Methods of forest management, to find acceptance, must there conform as closely as possible to existing methods of lumbering. Rules of marked simplicity, the observance of which will yet secure the safety of the forest, must open the way for more refined methods in the future.

For the present a periodic or irregular yield, temporary means of transport, constantly changing crews and an almost total ignorance of the silvics of all but a few of the most important trees - all combine to enforce the simplest silvicultural treatment and the utmost concentration of purpose on the two main objects of forestry, which are the production of a net revenue and the perpetuation of the forest. Such concentration has been followed in practice by complete success.

The forests with which the American forester deals are rich in species,usually endowed with abundant powers of reproduction and, over a large part of their range, greatly dependent for their composition and general character upon the action of forest fires. Of the commercially valuable trees there may be said to be, in round numbers, a hundred out of a total forest flora of about 500 species, but many trees not yet of importance in the lumber trade will become so hereafter, as has already happened in many cases.

The attention of the forester must usually be concentrated upon the growth and reproduction of a single species and never of more than a very few. Thus the silvicultural problems which must be solved in the practice of forestry in America are fortunately less complicated than the presence of so many kinds of trees in forests of such diverse types would naturally seem to indicate.

The forest fire problem is one of the most difficult with which the American forester has to deal. It is probable that forest fires have had more to do with the character and distribution of forests in America than any other factor except rainfall. With an annual range over thousands of square miles, in many portions of the United States they occur regularly year after year on the same ground. Trees whose thick bark or abundant seeding gives them peculiar powers of resistance, frequently owe their exclusive possessions of vast areas purely to the action of fire.

On the economic side fire is equally influential. The probability, or often the practical certainty, of fire after the first cut, commonly determines lumbermen to leave no merchantable tree standing. Forest fires are thus the most effective barriers to the introduction of forestry.

Excessive taxation of timber land is another of almost equal effect. Because of it lumbermen hasten to cut and afterwards often to abandon, lands which they cannot afford to hold. This evil, which only the progress of public sentiment can control, is especially prevalent in certain portions of the white pine belt.

Forest Associations

Public sentiment in favour of the protection of forests is now widespread and increasingly effective throughout the United States. As the general understanding of the objects and methods of forestry becomes clearer, the tendency, formerly very marked, to confound ornamental tree planting and botanical matters with forestry proper is rapidly growing less. At the same time, the number and activity of associations dealing with forest matters is increasing with notable rapidity. There are now about thirty such associations in the United States.

One of these, the Society of American Foresters, is composed exclusively of professional foresters. The American Forestry Association is the oldest and largest.

It has been influential in preparing the ground work of popular interest in forestry and especially in advocating and securing the adoption of the federal forest reservation policy, the most important step yet taken by the national government. It publishes as its organ a monthly magazine called *Forestry and Irrigation.* The Pennsylvania Forestry Association has been instrumental in placing that state in the forefront of forest progress. Its organ is a bi-monthly publication called *Forest Leaves.*

Other states which have associations or societies of special influence in forest matters are California, Massachusetts, Minnesota, Colorado, New Hampshire, Georgia and Oregon.

Arbor Day, instituted in Nebraska in 1872 as a day for shade-tree planting by farmers who had settled on the treeless prairies, has been taken up as a means of interesting school children in the planting of trees and has spread until it is now observed in every state and territory. It continues to serve an admirable purpose.

Lumbering

All the operations of the lumber trade in the United States are controlled and to no small degree determined, by the peculiar unit of measure which has been adopted. This unit, the boardfoot, is generally defined as a board one foot long, one foot wide and one inch thick, but in reality it is equivalent to 14 4 cub. in. of manufactured lumber in any form. To purchase logs by this measure one must first know about what each log will yield in one-inch boards.

For this purpose a scale or table is used, which gives the contents of logs of various diameters and lengths in board feet. Under such a standard the purchaser pays for nothing but the saleable lumber in each log, the inevitable waste in slabs and sawdust costing him nothing.

In addition to this amount, an immense quantity of wood is used each year for fuel, posts and other domestic purposes and the total annual consumption is not less than 20 billion cub. ft. The years 1890 to 1906 were marked by rapid changes in the rank of the important timber trees with reference to the amount of timber cut and a shifting of the important centres of production.

Among coniferous trees, white pine has yielded successively to yellow pine and Douglas fir, while the scene of greatest activity has shifted from the Northern forest to the.

Southern and from there is rapidly shifting to the Pacific Coast. The total cut of coniferous lumber has increased steadily, but that of the hardwoods is falling off and in 1906 it was 15 per cent less than in 1899, while inferior hardwoods are gradually assuming more and more importance and the scene of greatest activity has passed from the middle west to the south and the Appalachian region. *Conifers. The* coniferous supply of the country is derived from four forest regions:

- The Northern forest;
- The Southern forest;
- The Pacific Coast forest;
- The Rocky Mountain forest.

The Northern forest was long the chief source of the coniferous lumber production in the United States. The principal timber tree of this region is the white pine, usually known in Europe as the Weymouth pine. It has an average height when mature of i to ft., with a diamcter a little less than 3 ft., but the virgin timber is approaching exhaustion. White pine was one of the first trees to be cut extensively in the United States and Maine, the pine tree state, was at first the centre of production. In 1851 the cut of white pine on the Penobscot river was 144 million ft., that of spruce 14 million and of hemlock 11 million.

Thirty years later the pine cut had sunk to 23 million, spruce had risen to 118 million and hemlock had passed pine by a million feet. Meanwhile, the centre of production had passed from the north woods to the Lake States and for many years this region was the scene of the most vigorous lumbering activity in the world.

Woods waste includes tops, stumps, cull logs and butts, but does not include defective trees left or trees used for road purposes. Mill waste includes bark, kerf, slabs and edgings.

The spruce (*Picea rubens*) is used chiefly for lumber, but it is in large and increasing demand in the manufacture of paper pulp. For the latter purpose hemlock, poplar and several other woods are also employed, but on a smaller scale. The total consumption of wood for paper in the United States for 1906 was 3,660,000 cords, of which 2,500,000 was spruce. Of this, however, 720,000 cords were imported from Canada.

- The chief product of the Southern forest is the yellow pine. This is the collective term for the longleaf, shortleaf, loblolly and Cuban pines. Of these the longleaf pine, called pitch-pine in Europe, is the most important. Its timber is probably superior in strength and durability to that of any other member of the genus *Pinus* and in addition to its value as a timber tree it is the source of naval stores in the United States. The average size of the mature longleaf pine is 90 ft. in height and 20 in. in diameter. Shortleaf and loblolly are other important members of this group. Their wood very closely resembles that of the longleaf pine and is often difficult to distinguish from it. The trees are also of about the same size and height. Loblolly is, however, of more rapid growth. The total cut of yellow pine in 1906 was 11,661,000,000 board ft.; it has perhaps not yet reached its maximum, but is certainly near it.
- Another important coniferous tree of the Southern forest is the bald cypress, which grows in the swamps. The cut in 1906 was 839,000,000 board ft., a gain of 69 per cent over 1899.
- But the great supply of coniferous timber is now on the Pacific Coast. The Douglas fir, also known as Douglas spruce, red fir and Oregon pine, is the foremost tree in Oregon and Washington and the redwood in California. When mature the Douglas fir averages 200 ft. in height and 4 ft. in diameter and the redwood 225 ft. in height and 8 ft. in diameter. Other important trees of the Pacific Coast are sugar pine, western red cedar, western larch, Sitka spruce, western hemlock and western yellow pine.
- Logging on the Pacific Coast is characterized by the use of powerful machinery and by extreme skill in handling enormous weights. This is especially true in California, where the logs of redwood and of the big tree are often more than f ft. in diameter. Logging is usually done by wire cables operated by donkey-engines. The journey to the mill is usually by rail. The mills are often of great size, built on piles over tide water and so arranged that their product is delivered directly from the saws and dry kilns to vessels moored alongside. The products of the Pacific Coast forest make their way over land to the markets of the central and eastern states and into foreign markets. Among the lumber-producing states, Washington has in seven years jumped from fifth place to first and its output has increased from 1,428,000,000 board ft. in 1899 to 4,305,000,000 ft. in 1906. Oregon and California have increased their output from 734,000,000 each in 1899 to 1,605,000,000 and 1,349,000,000 ft. respectively in 1906. Of the total output of these three states

(7,259,000,000 ft.) 4,880,000,000 ft. is Douglas fir and 660,000,000 redwood.

- The important lumber trees of the Rocky Mountain forest are the western yellow pine, the lodgepole pine, the Douglas fir and the Engelmann spruce. The Douglas fir, here extremely variable in size and value, reaches in this region average dimensions of perhaps 80 ft. in height by 2 ft. in diameter, the western yellow pine 90 ft. by 3 ft. and the Engelmann spruce 60 ft. by 2 ft. Mining, railroad and domestic uses chiefly absorb the annual timber product, which is considerable in quantity and of vast importance to the local population.

2

Agroforestry System

Agroforestry is an activity that combines production on the same plot of land, from annual agricultural activities (such as crops and pasture) and from delayed long-term production by trees (for example timber and services). This is obtained either by planting trees on agricultural land or by cropping (for example after thinning) on forested land. Plots that combine arable intercrops with forestry trees are silvoarable plots, while wooded plots with pasture under the tree canopy are known as silvopastoral plots.

THE ADVANTAGES OF AGROFORESTRY

Agroforestry provides a different land use option, compared with traditional arable and forestry systems. It makes use of the complementarity between trees and crops, so that the available resources can be more effectively exploited. It is a practice that respects the environment and has an obvious landscape benefit. Efficient, modern versions of agroforestry have been developed, that are adapted to the constraints imposed by mechanisation. The agroforestry plot remains productive for the farmer and generates continuous revenue, which is not the case when arable land is exclusively reforested. Agroforestry allows for the diversification of farm activity and makes better use of environmental resources. Agroforestry has interesting advantages from three different perspectives.

From the Arable Perspective

Diversification of the activities of arable farmers, with the building-up of an inheritance of valuable trees, without disrupting the revenue from those plots which have been planted. Protection of intercrops and animals by the trees, which have a windbreak effect, providing shelter from the sun, from the rain, from the wind, holding the soil in place, and stimulating soil microfauna and microflora

Recovery of some of the leached or drained nutrients by the deep roots of the trees; enrichment of the soil organic matter by tree litter and by the dead roots of the trees. Possibility of combining the interest of the owner (for an inheritance of wood) and the farm (for access to cultivated land). Possible remuneration for the arable farmer for looking after the trees

An alternative to full reforestation of arable land, permitting the continuation of arable activity on land whose arable potential is therefore conserved. The tree component can be reversed, the plot stays "clean" (free from scrub) and is easy to destump when the trees are clear felled (the stumps are in lines and few in number). In silvopastoral plots, fodder units can be available at different dates compared to full cropped plots, extending the grazing calender.

From the Forestry Perspective

Acceleration of the diameter growth of the trees by wide spacing (+80 per cent over 6 years in the majority of the experimental plantations). Reduction of the capital cost of the plantation, by reducing the number of trees planted with no commercial future. A large reduction in the maintenance costs of the plantation, due to the presence of the intercrops. Improvement in the quality of wood produced (wide regular rings, suited to the needs of industry), because the trees are not subjected to cycles of competition and thinning.

Guaranteed follow-up and tree care due to the arable intercropping activity. In particular, protection against the risk of fire in susceptible areas, with pastoralism or with intercrops like vine or winter cereals (clear bare ground in summer after stubble ploughing). Agroforestry plantations on arable land allow the development of a quality wood resource that complements, rather then competes with, the products from traditionally exploited forests. It is especially important to produce wood that can substitute for tropical sawlogs, which will soon decline in availability and quality. The areas concerned will remain small in terms of their absolute value, but the production of wood from them could become a critical input to the European wood supply network. Tree species that are little used in forestry, but are of high value, could be grown in agroforestry systems: service trees, pear trees, common sorbs, walnut trees, wild cherry trees, maple trees, tulip trees, paulownias, etc...

From the Environmental Perspective

Improvement to the development of natural resources: the total wood and arable production from an agroforestry plot is greater than the separate production obtained by an arable-forest separate cropping pattern on the same area of land. This effect results from the stimulation of complementarity between trees and crops on agroforestry plots. Thus, weeds, which are spontaneously present in young forestry plantations are replaced by harvested crops or pasture; maintenance is less costly and environmental resources are better used. Better control of cultivated areas of land: by substituting for arable plots, the agroforestry plots contributes to diminishing the cultivated area of land. The intensification of environmental resource use by agroforestry systems is not resulting in more crop products.

Creation of original landscapes that are attractive, open and favour recreational activities. Agroforestry plots have a truly innovative landscaping potential, and would improve the public image of farmers to society. This will be particularly the case in very sparsely wooded areas, where plots are developed by planting arable land, and in very heavily wooded areas, where plots are developed by thinning the existing forest.

Counteract the greenhouse effect: constitution of an effective system for carbon sequestration, by combining the maintenance of the stock of organic material in the soil (the case especially with meadows), and the superimposition of a net fixing wooded layer. Protection of soil and water, in particular in sensitive areas. Improvement of biodiversity, especially by the abundance of "edge effects". This in particular, permits a synergistic improvement, by favouring the habitat of game. The integrated protection of crops by their association with trees, chosen to stimulate the hyperparasite (parasites of parasites) population of crops, is a promising way forwards. These favourable characteristics are as coherent with the many objectives of the laws guiding agriculture and forestry, as they are with the directing principles of the Common Agricultural Policy.

IMPORTANCE OF AGROFORESTRY

Agroforestry provides goods and services to the farmer. The main service roles include soil conservation, creation of a favourable microclimate, and societal benefits through the multitude of traditional roles that trees play in many civilizations.

Other service roles include nitrogen fixation (although nitrogen-fixing plants do not enrich the soil if most of the fixed N is removed from the site, as in the case of fodder); increased water retention capacity of the soil due to frequent loosening of the topsoil layer; and substantial reduction of weed competition when crops are tended. There are several products from multipurpose trees, and these can be used for subsistence or income generation, helping farmers to diversify production and so reduce risk. Agroforestry practices seek to address the problem of deforestation and to achieve a positive ecological interaction between the components of the system and the environment (especially the microclimate). At present, the problem of deforestation is intensifying, with trees being cut down at a faster rate than they are being replaced.

Moreover, the demand for fuelwood and other tree products is rising as population increases. Deforestation may also lead to loss of habitat for wildlife, and eventually to their extinction and hence loss of national heritage; to shortages of forest products such as timber, poles, firewood, charcoal and medicines; and ultimately to environmental degradation and erosion of genetic resources. In Mbale decreased fodder availability due to restricted access to forests led to a decrease in number of animals owned by households.

FARMING PRACTICES

Farming practices have to change if the land is to continue producing food for the ever-increasing population, for example continuous cultivation of the same fields leads to a decline in the productivity of the land and monoculture farming, coupled with the use of chemical fertilizers and pesticides, has an adverse ecological effect. Agroforestry can prevent a decline in soil fertility and productivity, which is an added motivation for practising it. Furthermore, the productivity of many tree products is superior on farms, as opposed to natural wooded land, because of farmer management.

Trees and shrubs are an important source of nutrition for livestock. Fodder trees (browse or top feed), are an effective insurance against seasonal feed shortages. They are used to supplement the quantity and quality of basal feeds such as grasses and crop residues. Fodder trees are less affected by dry conditions, owing to their extensive root systems and longer life spans. Incorporation of trees and shrubs for fodder in smallholder farming and intensive systems is gaining ground in Uganda.

Use of Agroforestry Systems

Calliandra has been introduced in farming systems as a multi-purpose shrub. There are many reasons for growing it on one's farm. These include 1) fuelwood, 2) stabilising soils and water conservation structures, 3) soil fertility improvement, 4) fodder supplement for ruminant livestock, 5) secondary production (bee forage, stakes for climbing beans and tomatoes), and 6) income generation through sale of wood, fodder and seed. Calliandra also provides several secondary products. It flowers prolifically and is a favoured source of nectar for honeybees. The honey produced is considered to be of high quality, with a bittersweet flavour. Processed into leaf meal, it has shown promise as a protein source for livestock.

Importance of Agroforestry Systems on Livelihoods

The economic viability of agroforestry systems has to be considered with regard to the benefits they can provide to farmers in Uganda. Calliandra can increase nutrient availability, improve crop output or, as fodder, increase milk production and so contribute to food security. Food insecurity is of major concern to Ugandans, as it is an important indicator of poverty. Smallholder production of food crops has been declining. Uganda has a high rate of infant stunting and mortality. Food insecurity, and particularly the inability of the smallholder sector to maintain or increase levels of food output, has been attributed to serious shortages of arable land, the low levels of technology practised by most smallholders and weaknesses in the delivery of agricultural and forestry services. Incorporation of Calliandra in the farming system can contribute towards reversal of these negative trends through its many uses in the farming system.

Use as a Fodder

Calliandra leaf material contains 17-28 per cent crude protein (dry matter basis) and can be used for dairy feed supplementation. Its use as a supplement to basal feeds, which in Uganda consist mainly of Napier and other grasses and crop residues, increases the overall protein content of the fodder ration and thereby increases milk production. Calliandra improves milk production of both dairy cattle and dairy goats.

In on-farm experiments with dairy cows under East African conditions, 3 kg of fresh Calliandra leaves provided a similar response in milk yield and butter per cent to 1 kg of commercial concentrate, increasing milk production by roughly 0.75 litre under farm conditions, although response varied depending on such factors as the cow's health and the quality of its basal diet. However, in a separate on-station experiment with sheep and goats, when concentrate or Calliandra was used in the same relative proportions as a supplement to Napier grass (without additional concentrate), the concentrate gave significantly better animal performance than Calliandra.

A cow needs to be fed roughly 6 kg of fresh leaves every day, while a goat needs about 0.7 kg. Increased milk production improves the nutritional status of households by increasing the proportion of milk in the diet. For those households that sell milk, income increases; and if prices become lower as a result of increased supply, poorer households also become able to afford to buy milk. Calliandra can also be fed to other types of livestock such as sheep, rabbits and chicken, which are more affordable to poorer households (although care must be taken with non-ruminants to feed Calliandra in small amounts, to avoid problems with toxicity due to tannins). Other feeds with high levels of condensed tannins are known to have anthelmintic (de-worming) properties.

To harvest 6 kg of fresh leaves every day on a sustained yield one needs to plant about 500 Calliandra trees at a spacing of 0.5 m (1.5 ft), making 250 m (800 ft) of hedge. Tuwei *et al* (2003) reported annual forage (leaf biomass production) yields of Calliandra of 13-15 t/ha/year (DM) in Embu, Kenya. The economic benefits of supplementation with Calliandra. In the first year, a small investment by farmers was required in planting labour and purchase of seedlings. This amounted to US $ 7.00 for the 500 shrubs needed. Supplementary feeding of 6 kg of fresh Calliandra fodder on a daily basis to a diary cow through a 10 month lactation period gave a net benefit of up to Ug shs 230,100 (US $ 122.32) per cow per year, beginning in the second year. This is as a result of increased milk production, assuming an increase of 10 per cent over base milk yields. Feeding 2 kg dry Calliandra day^{-1} as a supplement throughout the lactation period increased milk production by about 450 kg yr^{-1}.

BUDGET ECONOMIC ANALYSIS

In the same study, partial budget economic analysis showed that by feeding Calliandra as a substitute, the farmer saved money that would otherwise have been spent on buying and transporting 730 kg of dairy meal during the year. In the partial budget, assessing Calliandra as a substitute for dairy meal, incremental benefits per year after the first year were over 14 times higher than incremental costs. The net benefits per cow per year during years 2-5 were Ug shs 266,500 (US $ 141.68). In other words using Calliandra increased farmers' annual income by about US $ 120-142 per cow per year. These figures are based on the substitution rates (of Calliandra for concentrate) determined by the study of Paterson *et al.* (1996). Tree legume leaves are commonly fed to improve the utilisation of low quality grasses and crop residues in tropical areas. For example, goats in Zambia maintained on poor quality pasture hay lost weight (-20g/day), but gained weight (24g/day) when provided with supplements of Calliandra (140g DM/ day). Similarly, Palmer and Ibrahim (1996) found that increasing levels (0 to 35 per cent DM) of fresh Calliandra leaves in a diet of low quality hay increased the live weight gain of sheep from -27g to 52g/day.

Studies in Australia and Indonesia in the early 1990s suggested that forage of Calliandra had higher nutritive value when fed fresh. When leaf material of Calliandra was fed fresh to sheep, voluntary intake was 59g dry matter/kg $W^{0.75}$ ($W^{0.75}$ is the metabolic weight of the animal), but for dried leaves it was only 37g dry matter/kg $W^{0.75}$. More recent research, however, suggests that there is usually no significant negative effect of drying on animal production.

Positive Effects on Animal Nutrition

For non-ruminants, the high tannin content of Calliandra results in less positive effects on animal nutrition than is achieved with ruminants. Supplementation of poultry feed with Calliandra tends to increase feed intake and reduce egg production, although at low levels of supplementation yolk colour is improved. Calliandra is a source of carotene, which is used to maintain yolk colour in commercially produced eggs. Inclusion rates of Calliandra in poultry diet greater than 5 per cent decreased the number of eggs produced per chicken. Yolk colour was markedly improved at 5 per cent level and it was concluded that 5 per cent Calliandra leaves was the maximum level of inclusion consistent with the maintenance of production.

SCIENTIFIC APPROACH TO AGROFORESTRY

A scientific approach to agroforestry has only been developed in the past twenty years, with technologies intended to address land use conflicts and/or developmental and environmental problems. All agroforestry systems including those with Calliandra have benefits and opportunities as well as costs and constraints. In order to demonstrate to decision-makers the possible repercussions or trade-offs that result from alternative courses of action in implementing one agroforestry system over the other (or not implementing it at all), economic analysis is used. Such economic examination of alternative agroforestry systems can assist decision-making by offering a common monetary standard of measurement between alternatives. In the case of the individual farmer, economic analysis can help to ascertain whether or not agroforestry can increase productivity and farm income, and improve livelihoods in relation to traditional land-use agricultural activities. From a macro-economic perspective, economic analysis can show the expected economic outcomes of agroforestry interventions, to determine whether the net contribution to society can justify the expenditures incurred.

WIDE RANGE OF CLIMATIC AND SOIL CONDITIONS

Calliandra tolerates a wide range of climatic and soil conditions, but is affected negatively by frost and acidic soils. The plant is not tolerant of low oxygen tensions in waterlogged situations and does not grow well on poorly drained calcareous soils. Calliandra is a fast-growing tree, which can provide economic and environmental benefits in a short time (nine months to two years) making it easily acceptable to farmers. It can be grown in different niches on the farm: farmers in Uganda prefer to plant it on boundaries and soil conservation structures.

Good management reduces its competition with other crops for above- and below-ground resources. Calliandra has been widely promoted in Uganda by various organisations such as ICRAF-FORRI (NARO) since 1988, and by NGOs such as Heifer Project International (HPI) and its intermediaries operating in 25 districts, Send-a-Cow (23 districts), Africa 2000, Vi-Agroforestry Project (in 2 districts, since 1992), World Vision (22 districts with 1-3 sub-counties in each district) and the Uganda Land Management Project (in four districts). The various organisations cover up to 80 per cent of the country where there are farming households, however the proportion of farmers reached varies widely in the districts of operation, from three to fifteen percent.

The ICRAF-FORRI Agroforestry Research and Development Project had its initial focus in Kabale, in the Kigezi Highlands, and later expanded to Mukono District on the Lakeshore belt of Lake Victoria. The aim of the project is to improve livelihoods and ecosystem functions through the diversification of agroforestry systems. In Kabale and Mukono, farmers use Calliandra fodder in zero grazing (cut-and-carry) systems. It is estimated that 700,000 Calliandra seedlings have been planted every season since 1998 in Kabale District.

Calliandra is also widely promoted in other parts of Uganda by other agencies. For example, Vi-Agroforestry Project has been distributing seed to farmers in Masaka and Rakai Districts for the last ten years: in 2001 alone, a total of 1,574 kg was distributed while in 2002 it was 1,197 kg and in 2003 the total amount distributed increased to 2,309 kg (information from Vi organisation records and staff). In Uganda, there are regular feed shortages and droughts and therefore under such conditions subsistence feeding, mainly poor quality grasses and crop residues results in reduced live weight and low productivity of livestock. The severe nutritional limitations are further associated with delayed age at first calving, increased interval between parturitions, increased non-productive life of animals and herd wastage. Therefore, during times of feed shortages, foliage from shrubs and trees such as Calliandra have a great value as they supply dietary nitrogen.

Problem Statement

Many development projects are financed with public funds and a significant proportion fail to achieve their main objectives. It is therefore essential to assess the potential and actual impacts of research and development initiatives. The great majority of agroforestry research to date has concentrated on the biological and physical factors that affect productivity rather than the economic and environmental value of agroforestry. There is a serious lack of reliable information, based on actual farm conditions, about the economic benefits and costs inherent in many agroforestry systems. The cost of production of Calliandra seedlings and the costs and benefits of its use as fodder for supplementation and substitution, as well as other impacts experienced under farm conditions, were the main focus of this study.

Although a number of farmers have adopted Calliandra in Kabale, the enormous potential benefits of Calliandra and other trees for smallholders still remain largely unrealised in both the study districts (Kabale and Mukono) due to the low numbers of trees planted per farmer, for example, in Kabale, adopting farmers have planted an average of 320 trees, yet the recommended number for feeding a dairy cow throughout the year is 500 trees.

Secondly, there is lack of knowledge among farmers on how to use and manage the species. There are also many other outstanding questions. For example, who depends on planting of Calliandra? Are they rich or poor? Male or female? How does the use of Calliandra in agroforestry systems affect the incomes and welfare of farmers, especially the poor? The study was an attempt to answer these socio-economic questions and to quantify the actual, and determine the potential, socio-economic impact of Calliandra on farmers' household welfare in Mukono and Kabale Districts.

SEVERAL IMPORTANT OF JUSTIFICATIONS

There are several important justifications for this study:

1. Several of the agroforestry systems that have been tested using on-farm trials in Mukono and Kabale have shown positive results in terms of biophysical aspects and productivity increases at the plot level. However, little is known about the socio-economic impacts at the household level. For instance, observations made elsewhere (in Masaka and Rakai Districts) by Vi agroforestry staff and Mukasa (2003) have shown that farmers have complained about stiff Calliandra competition with bananas.
2. There is a lack of information on (a) the financial returns over time, (b) the labour requirements, and (c) the opportunity costs of land and crop yield for Calliandra-based agroforestry systems. Therefore, the profitability of agroforestry systems based on Calliandra, towards poverty alleviation, are unknown. Early adoption of technologies is mainly by well-off farmers, so the technology may have had little impact on the poorer farmers in the community. This study tried to fill this information gap by applying economic analysis both to primary data collected from farmers in the field, and to existing experimental data on the contribution of Calliandra.
3. There is little information on whether the incorporation of Calliandra in farming systems has contributed to environmental protection through planting, managing or changing land and other resource use practices. The study assessed the current and potential socio-economic and environmental impacts of Calliandra, as perceived by the farmers.
4. The generation of technologies that meet the needs of users is a necessary condition for the success of agricultural research. Only if farmers use the technologies developed through agricultural research, can research make an ultimate impact on the key goals of poverty alleviation, conservation of natural resources, and food security. Since not all outputs from research do indeed fully meet the needs of users, ongoing assessment of the use and impact of research provides crucial information. Moreover, even where innovations achieve widespread adoption and impact, they may not meet the needs of all farmers. In such cases, studies of adoption and impact can supply crucial information on the patterns of adoption, identifying who is by-passed by an innovation.
5. In Uganda, political support for agroforestry's contribution to the Plan for Modernisation of Agriculture (PMA) requires that the cost

effectiveness of various agroforestry technologies be demonstrated at the farm level. This study attempted to generate evidence that Calliandra-based agroforestry systems are cost effectiveness and that Calliandra has a role to play in the realisation of the goal of the PMA, *i.e.* 'eradication of poverty through profitable, sustainable and dynamic agricultural and agro-industrial sector'. Impact assessment is an essential requirement for achieving sustainable development. By integrating the economic, environmental and social dimensions of development into policy and project processes, it assists in advancing international and national development targets for long term sustainability.

Overall Objective

The overall objective of the study is to determine the potential and current socio-economic and farmers' perceived environmental impacts of Calliandra at the plot, household and community levels.

Specific Objectives

The specific objectives are:

- To determine the primary reasons for planting and primary benefits derived from Calliandra;
- To assess the cost of Calliandra seedling production and the costs, returns, and economic impact of Calliandra use as fodder;
- To establish farmers' own perceptions of the economic and environmental impacts of Calliandra.

Hypotheses Objectives

In line with the specific study objectives, the following hypotheses were made:

- Calliandra is a useful tree and meets the needs of farmers, although the primary reasons for its initial planting may be different from the primary uses at the time of utilisation;
- The economic benefits derived from establishment, management and use of Calliandra for fodder, outweigh the costs incurred in its establishment, management and use;
- There are more farmers who have experienced positive economic, social and environmental impacts of Calliandra than those who have experienced negative ones.

LIMITATIONS TO ENVIRONMENTAL IMPACTS

Environmental impacts could not be quantified. The data rely on farmers' perceptions, which were, however, cross-checked with secondary data from studies done elsewhere.

Analysis of Assumptions

It was deemed necessary to make assumptions about data collected and its analysis and the following assumptions were made:

- Selection of farmers for the study is representative of study area.
- The data are accurate and reliable.
- The tools used for data collection are accurate.

Scope of the Study

The study aims to assess the impact of Calliandra-based agroforestry systems, most of which have been tested by ICRAF/AFRENA in Kabale and Mukono Districts. In these districts, the focus is on sites at which there has been longer exposure to the Calliandra-based technologies.

In Mukono, the study focuses on fodder banks (the main use for which Calliandra has been promoted in this area), while in Kabale the focus is on hedgerows (contour hedges), fodder banks and rotational woodlots.

AGRO-FORESTRY; ADDING TREES TO FARMING SYSTEMS

It is relatively new to think of farming systems as including trees, or to plant trees with farming systems in mind. As recently as 1982, for example, a major study of farming systems made no mention of trees except as an "off-farm" source of fuel [1]. At the same time, most tree planting projects were assuming that farmers would grow woodlots on marginal land far from their crops.

Ideas are rapidly changing. A more recent study of African farming systems gives considerable attention to trees, especially in long sections devoted to soil conservation and the improvement of soil fertility 2. Simultaneously, social foresters have begun to give urgent attention to activities such as Niger's Maggia Valley Project, which planted windbreaks to protect farm La id from wind erosion. The convergence is not yet perfect: there are still foresters who seldom think of farms, and agronomists who seldom think of trees.

Nonetheless, there is increasing agreement on the need to grow trees within farming systems, or even to conceive of a wider range of plants and animal mix in order to create a sustainable, productive farming system: for example, to a maize monoculture might be added Leucaena leucocephala, bananas, sisal, groundnuts, small livestock, pigeon peas, cowpeas, mangos, yams, Cassia siamea, or Acacia albida. Properly arranged, these would serve to add nitrogen and humus to the soil, prevent erosion, fence off the garden, and provide food, fruit, poles and firewood for consumption and for sale. Since the soil would be automatically replenished with nutrients and organic material, such a system could continue indefinitely.

Conceived in this way, agro-forestry is a land use strategy that would: (a) increase the output of food and other products on a given unit of land; (b) ensure that increases in output were stable and sustainable; and (c) raise farmers' cash incomes. It is noteworthy that these are exactly the conditions necessary for greater food security in developing countries.

AGRO-FORESTRY TECHNIQUES

Many of today's agro-forestry techniques will be adaptations of traditional practices. However, non-traditional problems are raised by the decline of shifting cultivation, growing pressure on land, and rapid deforestation.

For example, there is new need for fast-growing, nitrogen-fixing tree species suitable for different ecological zones. In addition, ways must be found to compress larger numbers of trees, animals, and food crops on intensively managed farming systems. Intensive research is still required in this field.

Scattered Farm Trees

One agro-forestry intervention is simply to increase the number of trees or shrubs scattered among crops or pastures and along farm boundaries. These trees may have value greater than any crops they displace (as can be true of fruit trees). Alternatively, the trees may actually enhance the productivity of crops and pasturage by replenishing soils and helping to reduce erosion. For example, millet and sorghum may increase their yields by 50-100 per cent when planted directly under Acacia albida. Often, trees can be planted that will serve multiple purposes; Leucaena leucocephala, for example, can be used for poles, firewood, fodder and soil enrichment.

Scattering trees across the farm will not by itself meet all domestic needs for tree products or ensure sustainable agriculture. However, this kind of agro-forestry action is most likely to make immediate sense to farmers. Benefits are diverse and in line with felt needs. New species can be introduced cautiously and cried alongside proven ones. Risks are minimized. In a number of countries, this approach is now the principal way in which agro-forestry ideas are being introduced.

Improved Fallows

Another technique that can be introduced without making radical changes in the farming systems is the improvement of fallows.

This is required by the fact that fallow periods in many areas have been substantially shortened. As a result, only shallow-rooted grasses of little value to the soil establish themselves naturally before crops must again be planted. The usefulness of short fallows can be greatly improved by planting green manure crops such as Desmodium and Tephrosia spp., Lupinus mutabilis, Dolichos, Indogofera, etc. In addition to improving the soil, such crops may yield livestock feed or food for human consumption. It is worth noting that fallows can be improved without the use of "trees". Nonetheless, this kind of action falls directly within the scope of agro-forestry as defined here.

Alley Cropping

A much more complex agro-forestry technique is alley cropping, as developed by the International Institute for Tropical Agriculture (IITA). This involves the growing of crops between closely-spaced lines of trees or shrubs, usually legumes.

To prevent competition with crops for light and water, the trees are regularly cut back and are kept pruned during the growing season. Larger cuttings are available for firewood. Leaves and twigs are spread as mulch among the crops. This mulch replenishes the soil with humus and nutrients, as well as providing a barrier against erosion. Among the tree species found to work best in IITA's trial plots are Leucaena Leucocephala and Gliricidia sepium.

The details of alley cropping systems will vary. For example, herbicides may be applied (as IITA does), or weeding may be done by hand. Mulch may be spread before the crops are planted (as at IITA) or afterwards (as farmers have tried with mulching in northern Cameroon).

Trees may be cut at ground level (as at IITA), or just the branches may be lopped off (as farmers do on Mount Kenya, leaving a bare pole that does not shade the crops). The appropriate combination of techniques in a given area will depend en such factors as the availability of labour for mulching and weeding, the presence of pests Chat might be spread by mulching, specific crops to be grown, and the farmers' willingness to make large changes in farming practice.

Mulching

Mulch may also be used outside of alley cropping systems to protect or rehabilitate land. For example, the application of branch mulch helps restore nutrients to the soil, reduces erosion by breaking the impact of falling raindrops, and encourages water infiltration by slowing run-off and making the soil more porous.

Buffer Strips

Strips of vegetation planted along the contour lines of slops can serve as "buffers" against water erosion. Perennial grasses have long been used for this purpose in many countries. In more complicated systems, buffer strips may include grasses, fodder legumes, trees and shrubs. In such cases, the strip can yield mulch, forage, fruit and fuelwood, as well as controlling erosion.

Terracing

On gentle slopes, water erosion may be prevented through buffer strips and mulching. Where slopes are more severe, however, construction of terraces may also be necessary. This job will be made simpler if buffer strips are included in the form of dense vegetation planted along the risers and edges of each terrace. If this is done, a much smaller earth structure will be required to control run-off than would the case with conventional, "naked" terraces.

Windbreaks

Planting lines of trees to break the wind sweeping across open land can prevent erosion and slow the evaporation of moisture from the soil. Species can be chosen that will also yield fodder, poles or firewood. Individual farmers may act to protect land by planting trees on the windward boundaries of their farms. More ambitious schemes seek to control wind erosion over much larger areas.

For example, a CARE project in Niger's Maggia Valley established 250 km of windbreaks in seven years. These consisted of staggered, double rows of trees, primarily Azidirachta indica and Acacia scorpioides, planted across the valley at 100-metre intervals. Roughly 2,500 hectares of farmland were protected in this way. Initial indications are that millet production has been increased by 15 per cent above yields on comparable, unprotected land.

In contrast to most agro-forestry techniques, large-scale windbreak schemes necessarily involve whole communities. This raises delicate issues in terms of sacrifices (whose land will be pre-empted for planting the trees?) and benefits (whose land will be best protected from the wind, and who will get such by-products as fodder and firewood?). Social analysis will obviously play an important role in this kind of project.

Live Fencing

Many of the trees and crops in an agro-forestry system are vulnerable Co being eaten or trampled by livestock. Farmers will try to keep the animals away from the plants.

However, adequate supervision of the animals may be Coo time-consuming, and wire or metal fencing too expensive. To solve this problem, farmers can plant "live fences" in the form of tightly spaced lines of species such as Prosopis juliflora. Acacia scorpioides, or Euphorbia spp.

Woodlots

Farm woodlots have often been promoted to provide firewood, either for domestic consumption or for sale. As far as domestic consumption is concerned, it will sometimes be appropriate for a family to plant such a woodlot on marginal land. However, recent estimates suggest that household firewood needs may be harvested from as few as 50-100 trees. This is most likely to be true if the trees are scattered around a farm rather than packed into a woodlot, where yields per tree are lower. Any serious agro-forestry scheme would include several times that number of trees as part of windbreaks, buffer strips, or scattered farm plantings. Under these conditions, there would be little need for a separate woodlot to produce fuel.

The growing of firewood for sale will usually be less attractive. Almost everywhere in developing countries, farm-gate prices of firewood are extremely low. Under these conditions, any attempt at firewood farming will lose money, either in absolute terms or relative to alternative uses of the same land. Where woodlots are established, these are most likely to be for production of high-value outputs such as building poles or timber, although markets for such commodities are limited.

Project Components

The exact nature of any agro-forestry project will depend on the results of the rural appraisal outlined above. However, it is safe to assume Chat most projects will involve some form of research, demonstration, on-farm trials, extension, and monitoring and evaluation.

Research

Research must be done on the impact of including trees and other agro-forestry plants within farming systems. Both exotic and local agro-forestry species should be tried. Harvesting regimes should be designed to provide the products that people actually want, as determined by rural appraisals. Since many agro-forestry species grow slowly, research on these issues is a long-term endeavour. However, it may also be a long-term endeavour to convince large numbers of people to adopt complete agro-forestry systems. If research is begun immediately, results may therefore be known at just the point when people are looking seriously for proven techniques.

Demonstration

Even while research is still underway, it will be possible to demonstrate basic agro-forestry practices whose results are relatively well understood: for example, windbreaks; scattered plantings of Acacia albida, Leucaena spp., and fruit trees; live fencing; etc. Demonstration areas should incorporate crops, trees, and harvesting regimes that are consistent with local practices and desires. Management of these areas should be flexible, with species and techniques added or subtracted as they prove more or less convincing. An important part of this work would be to demonstrate agro-forestry plots alongside fields planted in traditional ways. This would allow farmers and other visitors to see quickly the differences made by agro-forestry techniques.

On-farm Trials

The ultimate test of an agro-forestry system is its performance on farmers' fields. At the farm level, results (and perceptions of results) may be very different from findings on research or demonstration plots. Agro-forestry projects will therefore need to sponsor on-farm trials of the most promising techniques. To be useful, these trials should as much as possible be conceived and evaluated by the farmers who are to carry them out. This implies the participation of farmers who are already fairly conscious of agro-forestry relationships. Results would be used to modify research and demonstration programmes, as well as to determine the agro-forestry message to be passed on by extension services.

Extension

The first extension job is to teach the extension agents themselves about interactions among trees, land, and food. Then, guidelines would be provided on how to evaluate farming systems, so that agents could guess what interventions might be appropriate under specific local conditions. Much information for this teaching programme would come from rural appraisals. Once the extension agents understood agro-forestry concepts, they could begin to introduce promising species and ideas. This would be done gradually, starting with straightforward techniques such as boundary plantings of fruit and leguminous pole trees. More complex techniques would be channelled through the system after being proved successful during on-farm trials.

Monitoring and Evaluation

Monitoring will be extremely complex, requiring measurement of yields (of both crops and trees) in very different places: research plots, demonstration areas, farmers' fields. These data are needed to assess the actual impact of various agro-forestry techniques. Equally important is to determine farmers' responses to agro-forestry. This should be judged by the number of farmers who become involved, and by what they say about their experiences.

National and Local Institutions

Each agro-forestry project would require a multi-disciplinary implementing group. A social scientist would be needed to conduct rural appraisals, to help design on-farm trials and extension messages, and to monitor project activities in terms of farmers' reactions and socio-economic results. The research programme would require joint effort by an agronomist, a forester, and an animal husbandry expert. An agro-forestry extension officer would supervise the communications support system, in collaboration with the national extension service. Additional technical and extension staff would be based at out-stations to run agro-forestry demonstrations and on-farm trials.

If the process is functioning properly, participation of farmers is guaranteed throughout. The original choice of research priorities and extension messages builds on the situation and attitudes of farmers. Farmer's' groups are identified or created to carry out on-farm trials and participate in other activities. Monitoring gives voice to farmers as the project proceeds. At the end, farmers are asked to help evaluate what has happened. Of course, this participation will not happen automatically; vigilance must be maintained to ensure that the farmer's point of view remains at the project's heart. Ideally, all of this work would be centred within the ministry already responsible for farmers and farming systems. It might prove desirable, therefore, to create an agro-forestry "cell" within the ministry of agriculture or rural development, made up of the kinds of professionals specified above.

In many countries, there exist non-governmental organisations (NGOs) that have close and continuing relationships with distinct groups of farmers. Some of these NGOs will already have experimented with tree-planting, generally as a means of meeting family needs for wood energy. Within an agro-forestry project, these NGOs could play an important role by carrying out demonstration activities and on-farm trials. However, close links would have to be maintained with the government, which presumably would retain primary responsibility for research and extension.

CHALLENGES FOR SUSTAINABLE ARID ZONE AGROFORESTRY

Agroforestry has been followed since time immemorial for sustainable livelihood in the arid zone, but in the recent past some changes have occurred, due to which the need is being felt to develop agroforestry systems, which may be able to meet the requirements of sustainable livelihood in the future.

HEAVY BIOTIC PRESSURE

Chalcolithic and Iron Age maps of the Indian subcontinent suggest that human occupation of the current Thar Desert region was thin during these earlier ages, and that populations depended upon hunting and limited pastoralism. From the analysis of available census records, the population of the Thar Desert region registered an increase of 490 per cent from 1901 to 1991. The present population of the Thar is about 40 million.

Decennial variation in population from 1901 to 1991 in the Thar showed a growth rate of 186 per cent (between 1901 and 1971) as compared with 132 per cent for the whole country. The population growth rate for the decades 1971-1981 and 1981-1991 were 36.7 per cent and 30.7 per cent, respectively. The population density is quite high compared with the global average of 6-8 persons/km^2 for arid zones. The major factor responsible for such phenomenal growth of population in the Thar is a wide gap between birth and death rates. Moreover, social values favour large families and positive sanctions for fertility outnumber negative ones.

Cropping on Marginal Lands

Of the total area of the Thar, only 1.3 per cent is now under forest cover. The net cultivated area increased by 54.1 per cent from 1915 to the early 1980s. This trend in cultivated area increase has further accelerated during the past decade (the 1990s), having increased by 11.7 per cent compared with the amount in the early 1980s. As a consequence of this, barren, cultivable and uncultivable wastelands, permanent pastures, and fallow land have declined substantially. There was a significant decline in grazing land area as marginal lands were brought under cultivation in recent years (the late 1990s). Because of low and erratic precipitation, restricted to a very short period, only one crop, the kharif (monsoon) season crop, is sown each year in most of the Thar.

The common crops are pearl millet, cluster bean, green gram, moth bean and sesame. In the arid zone of peninsular India, in addition to these crops, sorghum, maize pigeon pea and other coarse millets are also grown.

It has been observed that the agricultural holdings during late medieval times were only one fifth of those found today. By 1951, much of the potential agricultural land was brought under plow, and the trend since then in already established agricultural tracts has been intensification of farming, and the expansion of agriculture into new lands that may be considered environmentally marginal.

Mechanized field operations (use of tractors) not only increase soil erosion but also reduces the naturally regenerated population of various useful species.

Unsustainable Development Policies

With the advancement of science and technology, efforts have also been made to enhance production in the arid zone. Some of the technologies like sand dune stabilisation, wind-breaks and agri-horticulture have produced good results because they are developed in synergism with nature (ecosystem) and natives (people and vegetation).

But some other technologies imported from less environmentally stressed areas have proven to be unsustainable. For example use of tractors adversely affected the natural regeneration of native trees like Prosopis cineraria, and increases wind erosion. Irrigation with waters from the Indira Gandhi canal network has caused problems of water logging and, secondary salinization. Several exotic species either become invasive, like Prosopis juliflora, or didn't prove economically productive, like Faidherbia albida, jojaba etc. The overall impact of these imported technologies, along with high biotic pressure and a fragmented approach to desertification control, can now be observed in terms of loss of biodiversity, increasing desertification, and other adverse impacts. It is now imperative to think broadly and plan holistically for the future of arid zone agroforestry, the basis of sustainable livelihood in the arid zone. Throughout the arid zones, there is no dearth of problems, but rapidly increasing desertification (some call it land degradation) is a worldwide problem. Land degradation appears to be more of a descriptive term than desertification, as it better describes the process of wind and water erosion, soil salinization, and loss of vegetation cover. It helps in clarifying relationships among climate, topography, soil structure, land use, hydrology, and vegetation. Throughout the arid zones of the world, the major factor responsible for land degradation is soil degradation.

At the world scale, the GLASSOD database indicates that in arid zones, more than 60 per cent of soil degradation is caused by wind erosion, with human-induced water erosion accounting for 8 per cent of the phenomenon. At the global level, 349.6 million hectares of land in arid zones are characterised by a light to moderate degree of soil degradation, and 42.9 million hectares have strong to extreme soil degradation.

Besides wind and water erosion, over-grazing, high magnitude agricultural degradation activities, over exploitation of vegetation for domestic use, *e.g.* for fuel, and faulty land use are other factors responsible for ever increasing land degradation in arid zones.

The links between land or soil degradation and vegetation in arid zones are complex. Vegetation and soils display differing resilience to disturbance. Although in arid zones vegetation communities can be readily disturbed and degraded, recovery rates are very slow in comparison with those of other ecoclimatic conditions. Vegetation in arid zones is highty sensitive to natural climatic variability. Decreased production and an ability to lie dormant are two strategies displayed by arid zone plants for survival at times of moisture deficiency. Although an adequate database is not available, two useful relationships between soil degradation and vegetation can be determined. First, vegetation cover may help to indicate the vulnerability of a particular habitat to soil degradation, and second, vegetation cover or community type or plant species may indicate that soil degradation.

ARID ZONE AGROFORESTRY AND FOREST MANAGEMENT

As stated above, the prevalent form of forest management in arid lands is agroforestry, which is concerned with the management of trees and shrubs for conservation and sustainable development. Agroforestry in arid lands differs in many ways from the timber forestry practiced in moist tropical and other temperate ecosystems and in many cases its applications are broader in their scope. These applications include production of wood for fuel and poles; horticultural practices for fruit and nut production; range management for fodder and forage production; forestry to modify microclimate for improved agricultural crop production; and protection forestry on land susceptible to wind and water erosion.

Many rural people combine these applications of agroforestry with other land uses linked to their needs and social values. The value of forestry in the arid zone is difficult to evaluate in monetary terms. Arid zone agroforestry therefore needs to be understood broadly as management of trees and shrubs to improve the livelihood and quality of life for rural people in the arid environment. It is in this diverse context that policy developemnt must take place if sustainable development of forest resources in the arid zone is to be achieved.

Agroforestry is being followed traditionally in the arid zone to minimize the risk of drought and sustain life in climatic adversities. Several need-based systems have developed traditionally and some have been improved or incorporated in recent times.

ARID ZONE AGROFORESTRY

Drylands cover more than 60 per cent of the earth's surface. Arid zones are described as a part of the drylands, and have the most severe edapho-climatic conditions. The major distinguishing feature for defining and for planning for the arid zone is the low rainfall (below 500mm or Aridity Index <0.20) with more than 50 per cent inter-annual variability. This makes a great difference in terms of the nature of the ecosystem, the socio-economic environment and the challenges for sustainability. High wind and solar regimes further increase the effect of rainfall variability and the whole complex makes a fragile ecosystem in which small disturbances may cause great loss to the sustainability, which are sometimes irreversible. The hot arid zones of the world are economically and environmentally disadvantaged, with unique problems.

The ecosystems of these zones are highly fragile with large risks that cause severe impediments to development programmes. Of the total land area in the world, arid zones cover 18.8 per cent. The hyper-arid zones account for 22.3 per cent of the total arid zones. These arid zones are diverse in terms of climate, soils, vegetation, animals, and the lifestyles and activities of the people. Little but variability in rainfall and the presence of distinctive periods of drought are the characteristics of the arid tropics. Often, the terms drought and aridity are used incorrectly. A drought is a departure from average or normal conditions in which shortage of water adversely impacts on the functioning of ecosystems, and the resident population of people, whereas aridity refers to the average conditions of limited rainfall and water supplies, not to the departures there from. In general, arid zones are characterised by pastoralism and little farming but there are always exceptions.

Arid zones are found in almost all the continents of the world. In North America, arid zones are found in the southern US and Mexico, and in South America they are located in Brazil, Argentina, Chile, and in some other countries. In Africa, substantial parts of Saharan Africa, Ethiopia, and Namibia, are classed as arid environments. The hot arid regions of India lie between 24° and 29° N latitude, and 70° and 76° E longitude, covering an area of 31.70 million hectares, and involving seven states: Rajasthan, Gujarat, Punjab, Haryana, Andhra Pradesh, Karnataka, and Maharashtra. In total, 11.8 per cent of the country is under a hot arid environment. The arid regions of Rajasthan, Gujarat, Punjab, and Haryana together constitute the Great Indian Desert, better known as the Thar Desert, which accounts for 89.6 per cent of the total hot arid regions of India, and constitutes the principal hot arid zone of the country.

Table: Distribution of arid regions in different states of India.

State(s) *(10^6 hectares)*	*Area* *total*	*Percent of*
Rajasthan	*19.61*	*61.0*
Gujarat	*06.22*	*19.6*
Punjab and Haryana	*02.73*	*09.0*
Andhra Pradesh	*02.15*	*07.0*
Karnataka	*00.86*	*03.0*
Maharastra	*00.13*	*0.4*
Total	*31.70*	*100*

In arid zones, vegetation is typically sparse, and is comprised of perennial and annual grasses, other herbaceous plants, shrubs and small trees. The native plant species have adaptations that enable them to reproduce, grow and survive in the most inhospitable edapho-climatic conditions. Some plants have evolved special root systems, while others have unique leaf characteristics that allow them to withstand prolonged periods of drought. The number of tree species is very limited in arid zones, and in general, they are very slow growing due to limitations of environmental conditions, but nowhere in the world are they are so intricately associated with the life of human beings.

To evade or minimize the adverse affects of frequent droughts, the native peoples in arid zones have often developed production systems in which woody perennials have a very important role, both from a productivity as well as a resource conservation point of view. Their centuries-old experiences, under diverse socioeconomic and cultural conditions, were passed from generation to generation and people have established well-contained systems of production (silivipastoral/agrisilvicultural/

agrisilvipastoral/livestock husbandry) that accrue maximum benefits from their woody components. When arid zones have been looked at from the perspective of forestry development, the focus has been on the management of trees and shrubs that are either native to a particular arid zone, or have been introduced, especially for conservation purposes.

However, concern for trees and forests has increased in arid zones as it has in many other regions in recent years. This concern has been fuelled mostly by the disappearance of large areas of woodlands and trees from such fragile ecosystems in many countries of world. In the arid zone, closed forests are seldom available.

In India, traditional animal husbandry and agroforestry practices have been used to manage parklands, rangelands, reserved silvipastures near holy places, lay farming, and run-off farming (traditional watershed management). Trees are managed mainly for their non-wood forest products (NWFP) and their environmental services. Animals are an essential part of the production system. The whole system has been traditionally developed to spread the risk of drought in diversified components and efficient utilisation of scarcely available natural resources.

FUTURE DIRECTIONS FOR ARID ZONE AGROFORESTRY

Every ecosystem has some buffering capacity that enables it to remain unchanged for some time if it is overexploited.

While this applies to the arid zone, its fragile nature limits its buffering capacity. Thus, for a few decades the shortcomings of policy were not observed but as the buffering capacity is gradually being eroded, the indicators of unsustainability as a whole, *e.g.* decreasing natural resources, deforestation and desertification, become visible. For fragile ecosystems, like those that occur in arid zones, one thing needs to be recognised by policy makers and scientists, that is; sustainability of agroforestry in the arid zone depends on the degree of synergism with '3Ns' *i.e.* Nature, Native vegetation and Native people.

Some of the future directions for arid zone agroforestry need to focus on the evaluation, improvement and encouragement of sustainable traditional agroforestry systems, exploring the possibilities of natural regeneration/ rehabilitation of arid zone ecosystems, NWFPs as the basis of sustainable livelihood, rethinking of research priorities, and policy improvement. Evaluation, improvement and encouragement of sustainable traditional agroforestry systems that have been developed for centuries in harmony with nature. Agroforestry systems have been developed and used in this region since time immemorial. This may be because the inhabitants of this region found that no other option was sustainable aside from conserving and harvesting whatever nature is giving. The present forms of traditional agroforestry developed over the centuries through the experiments of earlier generations in all types of climatic variations. The validation of their sustainability is very difficult to establish by conducting short-term experiments lasting only 5-6 years. Thus, there is immense possibility for evaluation, improvement and encouragement of these traditional agroforestry systems.

Exploiting the possibilities of natural regeneration/rehabilitation of arid zone/arid zone agroforestry as an alternative to failed plantation/rehabilitation strategies. It is a mechanism of restoration or maintenance of ecosystems by nature that trees produce numerous seeds. A small part of this seed population is sufficient to replace the loss of plant population due to various factors. Human beings propagate the plants artificially to compensate for what they harvested form the natural vegetation or to get a higher quantity of the natural products. Those needed in large quantities have already been domesticated, *e.g.* food crops, fruit trees etc. Still a large number of natural species are wild or depend on natural regeneration. Trees in agroforestry may fall under this category. Natural regeneration as well as artificial tree establishing technologies works comparatively better in humid and irrigated areas. In the arid zone, due to climatological, physiological and sociological reasons artificial establishment has not achieved the required success. The need for domestication and technology development for sustainable harvesting and value addition for non-wood forest products (NWFPs).

Unlike the humid or sub-humid regions where trees are typically grown for timber, the trees are grown in this arid region mainly for non-wood forest products (NWFPs). In the arid zone closed forests are seldom available and agroforestry has been followed since time immemorial in which people conserve and manage whatever seeds of native tree species like Prosopis cineraria and Zizyphus numularia germinate in the crop field or on the common land. After the establishment of trees only annual lopping of branches is done which provides precious fodder during scarcity/drought. Taking the example of P. cineraria, this tree is found in this region at densities ranging from 40 to 120 trees/ha.

This we can say is the natural forest in common or private lands otherwise no dense forest is found as in the high rainfall areas. People lop the branches of this tree in the month of December and it becomes lush green in the month of May-June when no other fodder is available. Simultaneously, pods develop which are used as vegetables. Similarly other tree species provide fodder, pods, fruits etc. The interesting thing is that, during drought when annual crops fail, the yield of NWFPs does not decrease drastically, thus trees provide food security during drought. Besides some trees produce other products like gum, resins, edible/non-edible oils etc. which increase the income of the marginal farmers who are in the majority in this region.

The services provided by the trees in hot arid regions are equally important for moderating the climate. Cultivated in the form of shelterbelts, they reduce wind erosion and conserve soil and water.

In agroforestry systems, trees reduce the solar heat load on crops and also decrease wind speed and soil temperature, conserve soil moisture, increase biological activities in the soil and thereby improve the microenvironment for crops. The yield of crops grown under P. cineraria is generally higher than sole crops, knowing this fact agroforestry has been followed in this region for centuries. The edapho-climatic conditions in this region are so harsh that every tree improves the environment, it may be at micro level, yet its impact is visible. The other important NWFPs in these arid areas are plants with medicinal values. Due to climatic specialty there are medicinal plants which are either endemic to this region or their quality deteriorate if they are grow/found in higher rainfall areas.

Research Priorities: Considering the social and ecological demands, policy on agroforestry research also needs to be prioritized. Some of the possible priorities are-In-situ rainwater conservation for tree establishment and for agroforestry more generally.

Water is the most limiting factor for productivity and sustainability in the arid zone. As described earlier, the effect of bringing additional water through canals or ground water has been the cause of unsustainability in the ecosystem. Thus the only possibility remains the conservation of rainwater. Trees are the keystone of arid zone conservation and once trees are established the other conservation processes start automatically *e.g.* rainwater conservation, development of vegetation cover etc.. Further research is required on:

(i) Development of in-situ rainwater conservation techniques that require less labour and maintenance;
(ii) Development of plant-based water absorbing/retaining materials;
(iii) Initiatives for widespread adoption of rainwater conservation techniques;
(iv) Improvement in traditional water harvesting systems;
(v) Development of simplified windbreak establishment techniques.

Improvement of Policy: There are some important aspects on which improvement of policy is needed for creating a more congenial environment for development. These include:

Increased international funding for arid-zone research and development. The arid regions of the world in general suffer from a vicious cycle of low productivity, a low level of investment and as result, poverty. Greater priority is given at the international level for funding of projects and programmes related to timber forestry and rainforest management, perhaps because these activities contribute more to international trade. Very limited funds aimed at combating desertification are available to the arid lands. Forestry and agroforestry activities in arid regions no doubt contribute less to international trade but they provide substantially support to the livelihood of people in these climatically less favoured regions.

There is a need to rationalize these priorities and for the international community to give due importance to the non-tradable benefits of forestry/agroforestry, which are otherwise the basis of sustainability and livelihood. Suggested areas for increased investment in the arid zone are as follows:

(i) *Organic farming in virgin arid lands*. The philosophy of the traditional systems was to work in synergism with nature and use locally available natural resources. High input agriculture (HIA) on the other hand is only sustainable with assured input supply, *e.g.* irrigation, fertilizers etc, which is not always possible in drylands. Thus the chances of failure are much greater than in traditional systems during drought. Also due to less use of synthetic chemicals in the past, these lands are comparatively free from the residues of such chemicals. Hence organic farming, based on traditional systems and conservation of local natural resources can be one area for investment without much effort. Organic farming of NWFPs can be the first choice for investment.

(ii) *Eco-tourism*. Investment in eco-tourism can prove to be a large income generator in the arid lands. Increasing numbers of people are spending increasing amounts of time and money enjoying the unique resources of these arid land environments, *e.g.* spectacular scenic beauty, diverse wildlife population, relatively clean air and other amenities such as large open spaces. Investment in the forestry sector can be in amenity plantations, desert parks, buffer strips, roadside plantations and green belts can be economically as well as environmentally beneficial for the arid zone.

Exploring the potential of arid zone agroforestry as a tool for solving environmental problems. Conservation of Biodiversity: Biodiversity is the essential safety mechanism of any ecosystem and as the arid zone is considered to be fragile in nature the importance of conservation of biodiversity is further increased. Because of biotic interference and developmental activities, not only is the population of various species reduced but also their natural habitats have been destroyed. Thus the easiest way to save these species is ex-situ conservation with protection. This is possible through arid zone agroforestry, in which all types of plants *e.g.* trees, shrubs, climbers etc can be accommodated, combining conservation with domestication for increased production.

Desertification Control: Desertification is the result of exploitation of arid zone ecosystems beyond their carrying capacity. The carrying capacity of the arid lands is inherently low thus they are highly prone to desertification. Through soil and water conservation and sustainable management of vegetation, the carrying capacity of natural resources can be enhanced.

Various arid zone agroforestry models, *e.g.* wind breaks, sand dune stabilization, silvi-pasture, agri-horticulture etc. can be appropriate for soil and water conservation in arid lands, and the productivity of lands suffering form desertification can be enhanced and in this way desertification can be controlled. Arid lands can be considered simulated desertified drylands for planning of desertification control.

This can be important for preventive strategies for drylands rather than to control desertification after the process starts. Any technology or strategy that is found suitable to increase the productivity of arid lands, would give better results if applied to the desertification-prone drylands because they have a better environment than the arid lands. Assessing the potential of arid zones as carbon sinks.

In addition to their other benefits, trees on farms or in pastures also store carbon. Arid lands cover almost one forth of the earth's surface and any effort to increase tree populations substantially contributes to their value as carbon sinks. Due to their perennial nature and high root turnover, grasses can also prove to be a good carbon sink.

Silvipasture systems are prevalent in most of the arid zone and when both the tree and grass components are properly managed, an increase in net carbon storage can be achieved.

The arid lands are prone to desertification and trees and shrubs can be grown on these lands to prevent further soil degradation and to produce biofuel. These trees also enhance soil organic carbon content and thereby increase soil carbon storage. About 104 million ha land area in the world is affected by strong and extreme soil erosion.

FUNCTIONS OF AGROFORESTRY BIODIVERSITY

Throughout the world human communities have played a central role in shaping nature's diversity and its associated functions. Biological diversity has contributed in many ways to the development of human culture, and mankind has in turn influenced biological diversity.

As an example, gourds (*Lagenaria siceraria* and *Luffa spp.*) show tremendous varietal diversity and have been selected for a multitude of use and functions, including containers, pipes, scrubbers, floats, musical instruments, penis sheaths, ornaments and food. Plants and animals, both wild and cultivated, have been combined in complex and diverse agroecosystems in terrestrial and aquatic environments. At the broader landscape level, recent scientific evidence suggests that virtually every part of the globe-from boreal forests to the humid tropics-has been inhabited, modified and managed for millennia. Over time, human influence has shaped the expression of agricultural biodiversity at the genetic, species, ecosystem and landscape levels.

The present paper is a contribution to the FAO/Netherlands Conference on the "Multifunctional Character of Agriculture and Land-MFCAL" that will take place in Maastricht, The Netherlands, in September 1999. This background paper focuses on the relationships between agricultural biodiversity and the functions of agro-ecosystems at different spatial and temporal scales.

Selected examples are used to highlight the multiple functions of agricultural biodiversity and its links with rural livelihoods in a range of ecological and economic settings. The paper is divided into three parts.

Part one draws from the MFCAL approach, and in particular focuses on the contribution made by agricultural biodiversity to 1) food and livelihood security 2) production and environmental sustainability and 3) rural development. Part two identifies the forces which impact on agricultural biodiversity. Part three concludes by outlining some of the policy and institutional reforms needed to sustain agricultural biodiversity and agro-ecosystem functions.

THE MFCAL APPROACH

The MFCAL Approach provides the overall referential framework for this chapter. The approach recognises that the first and foremost role of agriculture remains the production of food. It stresses however, that agriculture and related land use activity can also deliver a wide range of non-food goods and services, influence the natural resource system, shape social and cultural systems and can contribute significantly to economic growth. The approach also focuses on the trade-offs and synergies that can exist between these different functions.

The first dimension of the multifunctional character of agriculture and related land use concerns food production and the contribution that this makes to food security.

Food security has been defined by FAO as a situation in which all people at all times have physical and economic access to sufficient, safe and nutritious food to meet their dietary needs and food preferences for an active and healthy life. There are three dimensions implicit in this definition: availability, stability and access.

Adequate food availability means that, on average, sufficient food supplies should be available to meet consumption needs. Stability refers to minimising the probability that, in difficult years or seasons, food consumption might fall below consumption requirements. Access to food draws attention to the fact that, even with bountiful supplies, many people still go hungry because they are poor and unable to produce or purchase the food they need. In addition if food needs are met through exploiting non-renewable natural resources or degrading the environment there is no guarantee of food security in the longer-term.

Current concerns for food security stem from both the unacceptability of current levels of food insecure people (at least 800million people) and the recognition that agriculture will have to feed an increasing human population, forecast to reach 8 000 million by 2020, of whom 6 700 million will be in developing countries. In most developing countries, the majority of the poor live in rural areas and depend on agriculture for their livelihoods. Expanding food production to feed this increasing population, while alleviating poverty through gainful employment in agriculture, is a formidable challenge.

In addition to food production and the vital contribution that it makes to food security, the MFCAL Approach recognises three further broad functions 1) environmental 2) economic 3) social.

The Environmental Function. Agriculture and related land use can have beneficial or harmful effects on the environment. The MFCAL approach can help to identify opportunities to optimise the linkages between agriculture and the biological and physical properties of the natural environment. The environmental function of the MFCAL Approach is relevant to a number of critical global environmental problems including biodiversity, climate change, desertification, water quality and availability, and pollution.

The Economic Function. Agriculture remains a principal force in sustaining the operation and growth of the whole economy, even in highly industrialised countries. Valuation of the various economic functions requires assessment of short, medium and long-term benefits. Important determinants of the economic function include the complexity and maturity of market development and the level of institutional development.

The Social Function. Both the maintenance and continuing dynamism of rural communities are basic to sustaining agro-ecology and improving the quality of life (and indeed, to assuring the very survival) of rural residents, particularly of the young. On another level, the capitalisation of local knowledge and the forging of relationships between local and external sources of expertise, information and advice are fundamental to the future of existing rural communities. Social viability includes maintenance of cultural heritage: for we know that in many instances, societies still identify strongly with their historical origins in agrarian communities and rural lifestyles.

AGROFORESTRY SYSTEM

The significance of agroforestry as a land use system has been highlighted extensively in recent years. Among the many favourable attributes of agroforestry is its suitability for degraded land. In India, the area of land that belongs to this category is considerable. Commonly known as wastelands, the total area of such land in the country is variously estimated at less than 70 million to more than 170 million ha (Radhakrishna, 2006). Even to allow the low end of this range to remain idle is ill-affordable to the country as the pressure on natural resources is immense.

A systematic approach is necessary to rehabilitate wastelands because they are a national wealth remaining underutilised. Agroforestry is a potential alternative suggested by many for the rehabilitation of wastelands. Their current fertility status is such that the probability of success of annual cropping systems is low whereas agroforestry has a reasonable chance to succeed. Along with the material harvests from the perennial component, soil fertility improvement is a benefit of agroforestry. Thus, fertility restoration is a key objective of agroforestry.

Meeting the energy needs of a rapidly-growing economy is another matter of apprehension for India. Issues associated with conventional petroleum energy are fluctuating global prices, depletion of reserves, dependence on imports and environmental pollution. Understandably, efforts have been intensified to identify alternative sources of energy. Biodiesel has emerged as one of the promising options and production of its raw material, oilseeds, is now gaining wide publicity. Although all oils can be used, edible oils as a source for biodiesel production have to be ruled out because they are required for cooking and food purposes. Therefore, non-edible oils are the premier raw material for biodiesel production in India. Seeds rich in these oils are mostly produced by perennial species. Hence these trees are referred to as Tree-Borne Oilseed Species (TBOS) and they produce Tree-Borne Oilseeds (TBOs).

Production of TBOs on wastelands meets the twin objectives of rehabilitation of a vital natural resource and obtaining an energy substitute. This is a highly appealing combination because it includes a number of ecological and economic benefits. Establishment of agroforestry systems with TBOS on vast areas of barren or scrub land as well as less fertile farm land will change the rural landscape and bring with it many environmental benefits.

Environmentally, the gains can be in the form of reduced erosion, improved micro-environment for intercrops, enhanced soil fertility, production of raw material for less polluting biofuel and by-products for various uses. The economic benefits being highlighted are increased rural income, reduction in expenses on petroleum imports and energy at lower cost.

Although there are definite advantages in promoting TBOS-based agroforestry on barren wastelands and marginal farmland, achieving the desired outcome is not easy. The major constraint, in this regard, is the lack of accurate information about TBOS, their cultivation practices, potential yields and income. In the absence of reliable sources on the topic, interested growers have to rely on whatever information that comes their way. Unfortunately, most promotional literature available at present tends to exaggerate the returns and understate the efforts required to succeed.

In the wake of the publicity jatropha and other TBOS are receiving, there is a proliferation in the number of stakeholders. Unfortunately, a majority of them are misled by promoters who are overenthusiastic or are taking advantage of the prevailing euphoria. At stake are the feasibility of agroforestry as a production system and the suitability of TBOS as a source of energy. Among the many issues to be addressed is the need for a systematic effort to develop and disseminate appropriate production packages so that an excellent opportunity is not lost.

MEANING OF AGRO-FORESTRY

Agro-forestry is an old concept. Trees, crops and animals have traditionally been raised together on small farms throughout the World. This concept first derived in the temperate zone due to the small family farms, as a result trees, crops and animals become separately managed on a large scale in modern agriculture and forestry. In India also we exploited our natural resources and adopted this sectorial policy.

In our country the functional allocation on land is 46.4 per cent for Agriculture and 22.7 per cent area for forestry is not sufficient for meeting the multi ferrous requirement of growing population for food, fodder and fuel and other raw materials. The only answer appears to be to integrate the land used for agriculture and forestry in such a way as to maximize production of foods and services for diverse requirements rural communities. zModern Agro-forestry establishes a symbiosis among agricultural crops tree species and livestock rising. In other words, these are complementary and beneficial to each other.

In short conventionally there had been separation between Agriculture and Silviculture. From immemorial on a limited scale a combination of food crops and forest crops had been adopted in land management by the farmers throughout the world, however due to steep rise in the demand for fuel wood and food, due to increase of population, present and early and urgent necessity to adopt the scientific approach on a large scale to the dual system of production or co-existence of forestry and farm council of research for FAO set Agroforestry in 1878 and it is a landmark in history in Agro-forestry. Growing of forest crops along with food grains in dual system has been extensively being adopted world over by the farmer.

Agro-forestry is defined as an efficient, integrated and sustainable land use system that combines Agricultural crops, Forest corps and/or Livestock together on the unit of farmland at the same time or in sequential manner.

In an Agroforestry there are both ecological and economical interactions between the various components.

Agro-forestry is collective name for land used systems involving trees combined with crops and/or animals on the same unit of land:

- It combines production of multiple outputs with protection of resources.
- It places emphasis on the use of multiple indigenous trees and shrubs.
- It is particularly suitable for low input conditions and fragile environment.
- It is structurally and functionally more complex than mono culture (single crop culture) the cycle of Agroforestry system is always more than one year.

OBJECTIVE OF AGRO-FORESTRY

- To manage land efficiently so that its productivity is increased and restored.
- To use available resources efficiently and economically
- To generate employment opportunities for rural peoples.
- To provide raw material for small cottage industries in rural areas.
- To raise the supply of fuel in the rural areas at convenient distance for consumer. In India 70 million tons of dried cow dung is used every year, which can be diverted for natural organic fertilizer moreover undue pressure is on traditional forest for obtaining fuel wood.
- Agro-forestry aims to raise the supply for small timber used by villages for agricultural implements, house construction and other domestic purposes. In this way Agroforestry can meet this requirement of the rural population and reduce pressure on forest.
- One of the main objectives of Agroforestry is to raise the production of food crops, legumes and tuber to meet the rapidly growing food requirements of the Indian population.
- Agroforestry aims at promoting production of, vegetables, pulses, milk and meat. Thus it can raise the Nutritional value of food, which is urgently, require for mankind in our country. Average Indian gets 2000 calories when 3000 calories require per day.
- Agro-forestry programme helps in obtaining an ecological balance in rural areas and thus it may be consider a matter of great significance for a country like India.
- Preservation of humidity in cultivable lands and check soil erosion. Increase productivity of land. In drought prone areas Agroforestry reduces insecurity of the agriculture; in such areas the dual system

of production of tree and grasses ensures stability with productivity of land.

- Supply of fodder for vast population of livestock. For proper feeding to livestock increase supply of fodder is urgently required. Large supply of milk and meat is achieved from livestock and poultry when fodder and feeding is proper.

CHOICE OF SPECIES FOR AGRO AND FARM FORESTRY

Following considerations mainly motivate the selection of species in social forestry:

- Rapid growth
- No competition with field crop.
- Fixation of Nitrogen
- Easy decomposition of litter or leaves
- Fast growth and easy establishment
- Ability to regenerate
- High yield of wood and fodder
- Multiple use of wood like Subabul
- Plant with deep tap root system
- Plant yielding small timber
- Easy in establishment and ability to coppice
- Capacity to grow under wide range of environment, soil types, rainfall etc.
- For recreation and shade plants with quick growing and give shade and pleasing colour like Spathodia, Gulmohar, Cassia, Jakaranda are selected.

KINDS OF LAND AND SITE FOR AGRO AND FARM FORESTRY

The following type of land and sites can be assigned for Agroforestry:

- Field boundaries
- Along with farm roads
- Along with nala sites
- Land on which cultivation is difficult
- Old fallow lands
- Cultivable wasteland
- Site of cattle shade, kitchen farm, farms of house etc.

BENEFITS FROM AGRO-FORESTRY

Combining trees with food crops on cropped farms yield certain environmental benefits such as:

- Reduction of pressure on Forrest.
- More efficient recycling of nutrients by deep rooted trees on the site.

- Better protection of ecological systems.
- Reduction of surface run-off, nutrient leaching and soil erosion.
- Improvement of microclimate, such as lowering of soils surface temperature and reduction of evaporation of soil moisture due to mulching and shading.
- Increment in soil fertility through addition and decomposition of litter fall.

Economical Benefits

Agro-forestry brings significant economic benefits to the farmers, the community, the region and the nation such as:

- Increment in outputs of food, fodder, fuel wood, timber and organic matter.
- Reduction in incidence of total crop failure.
- Increase in levels of farm incomes due to improved and sustained productivity.

Social Benefits

- Improvement in rural living standards from sustained employment and higher income.
- Improvement in nutrition and health due to increased quality and diversity of food.
- Provides stability to rural peoples.
- Ecological balance.
- Pollution reduction.

LIMITATIONS OF AGRO-FORESTRY

Agro-Forestry does have Certain Negative Aspects

- Possible competition of trees with food crops for space, sunlight, moisture and nutrient which may reduce crop yield.
- Damage to food crops during harvesting of trees.
- Potential of trees is serving as hosts to insects and birds.
- Rapid regeneration of profile trees may displace food crops and take over entire fields.

Through skilled management practices any or all these aspects can be controlled. For example, once it is known that trees complete with food crops and may reduce food yields, it is easy to adopt some of the following strategies.

- Select legume trees that have small or light crowns so that sunlight will reach the food crops.
- Select trees that are deep-rooted so that they will also absorb moisture and nutrients from the deeper subsoil.

- Space the trees further apart to reduce their competitive effect on the food crop.

SOCIO-ECONOMIC ASPECT OF AGROFORESTRY

- Requirement for more labour inputs, which may cause search at times in other farm activation.
- Competition between food or fore crops which could cause aggregate field, to be grown than those of single crop.
- Longer period required for trees to grow to mature and acquire an economic value.
- Resistance by farmers to displace food crops with trees especially where land is scarce.

CLASSIFICATION OF AGRO-FORESTRY SYSTEM

Different types of Agroforestry systems exist in different parts of the world.

These systems are highly diverse and complex in character and functions.

To evaluate understand and seek to improve them requires their classification into different categories.

Several criteria can be used in classifying them, but the most common include the system's structure, functions, and socio-economic scale of management and ecological spread.

According to Nair (1987), Agro-forestry systems can be classified according to following sets of criteria.

Structural Basis

Consider the composition of the components; specially refer including spatial admixture of the woody component, vertical stratification or the component mix and temporal arrangement of different components.

Functional Basis

This is based on the major function or role of the system; mainly of the woody components (This can be productive or protective).

Socio-Economic Basis

Consider the level of inputs or management (low input, high input) or intensity/scale or management and commercial goals.

Ecological Basis

Take into account the environmental conditions on the assumption that certain types of systems can be more appropriate for certain ecological conditions.

CLASSIFICATION OF STRUCTURAL BASIS AGRO-FORESTRY SYSTEM

In these systems the type of component and their arrangement are important. On the basis of structure, Agroforestry systems can be grouped into two categories:

- Nature of components
- Arrangement of components.

Nature of Components

- Agri-silvicultural Systems
- Silvipastoral Systems
- Agro Slivipastoral Systems
- Other Systems

Arrangement of Components

- Spatial Arrangement
- Temporal Arrangement

AGROFORESTRY IN INDIA

SCOPE OF AGRO-FORESTRY IN INDIA

There is tremendous scope for Agroforestry because India has achieved self-sufficiency in food production. Now its attention is becoming more focused on the ecological problems and shortage of fuel, fodder and other outputs as well as unemployment.

Agroforestry has vast scope in meeting this requirement through multipurpose tree species as:

- Large area is available in the form of farm boundaries, bunds, waste lands where this system can be adopted
- This system permits the growing suitable tree species in the field where most annual crops are growing well
- By growing trees and crops on Agricultural or forest land, Resources are utilized efficiently
- System has potential generate employment.
- Provides raw material for the cottage industries
- Helps in maintaining ecological balance
- Soil and water conservation, soil improvement.
- Helps in meeting various needs of growing population.

SYSTEMS

Agroforestry is a traditional practice among farmers in India. The practices adopted may not be well-defined systems like those evolved scientifically in recent times. Nevertheless, they too have followed the same principles and realised the benefits. A comprehensive description of all these practices and systems is found in the documentation of Tejwani (1994). There are many traditional agroforestry systems in practice in arid regions of India (Nandal and Narwal, 1994). Presence of *Acacia nilotica* and *Dalbergia sissoo* is very common in northern parts of the country. In the humid tropical areas along the south-western coast, systems which combine several perennial species are a common occurrence. The inclusion of several spice crops in these intensively-managed systems makes them highly remunerative.

TBOS have not been a key component of any traditional agroforestry system and often they may not have been established deliberately by farmers. The exception is jatropha, which is established by farmers as a live fence. This requires planting them in a row at close spacing. Other species like *Pongamia pinnata*, *Madhuca* species and *Calophyllum inophyllum* probably came up as wildlings and were retained by farmers. Their presence is usually limited to a few trees in the middle of crop fields or along the borders. Oilseed production is not the intention of retaining these trees and they are mostly there for service functions like providing shade and protection as fence.

Agroforestry technology development has been an active programme in the country for many years (NRCAF, 2004). In recent years, several agroforestry systems have been successfully introduced and accepted by farmers in India. Systems based on poplar (*Populus deltoides*) in the north-western region of the country are highly intensive and are suitable for irrigated conditions. An agri-horti-forestry system - where the component species are horticultural crops like mango and cashew, annuals as intercrops and multipurpose trees along the border of the farm - has gained acceptance in Maharashtra and Gujarat (Mahajan *et al.*, 2001). Additionally, there are many innovative farmers who have developed or modified existing agroforestry systems to suit local conditions. TBOS can fit into most of these systems, contributing positively towards the overall productivity and farm income.

LIMITATIONS

Wastelands belonging to farmers that can be brought under TBOS is available in plenty. The present production levels of some marginal farmlands are so low that farmers will be happy to try out alternative crops. Therefore, land availability for oilseed production is unlikely to be a constraint. The vital resource that will determine the area eventually brought under oilseed production is water. Although most TBOS are hardy, they will still require adequate water to produce satisfactory yields. Because priority allocation of available water would be for food production, oilseed production has to be concentrated on land where competition for water does not occur.

TREE-BORNE OILSEED SPECIES

Seeds of many tree species contain high levels of oil and their use for bioenergy generation has been a topic of interest for long (Raina, 1986). Those popularly known as TBOS are *Jatropha curcus* (jatropha or ratanjyot), *Pongamia pinnata* (pongamia or karanj), *Madhuca latifolia* and *M. indica* (mahua), *Calophyllum inophyllum* (undi), *Azadirachta indica* (neem) and *Simarouba glauca* (simarouba). Simarouba has been studied to standardise various aspects of it cultivation (Joshi and Joshi, 2004).

It is not a very familiar species in India. It produces edible oil and probably requires relatively better growing conditions compared to others in this group. Mahua oil is also edible and is used by tribal communities. Undi probably has the highest seed oil content among these species, but its major limitation appears to be the restricted environmental niche of sandy soil with humid environment. Moreover, mahua and undi are very slow-growing species and hence may not fit into an agroforestry system. Neem has recognition more for its pesticidal uses than seed oil. Thus, the list narrows down to jatropha, pongamia and neem. Jatropha appears to be the frontrunner at present.

There are several reasons for the greater interest in jatropha. It already has a domestication record as research on it has been carried out in India and other countries. Seed oil characteristics of jatropha are superior to others for biodiesel production. Besides its faster growing attribute when compared to most other TBOS, it also has the ideal size for agroforestry. Thus, jatropha has emerged as the premier TBOS. Unlike most other perennial species, jatropha has a shorter gestation period and regular seed harvests are possible within four years of establishment. Being a small tree with a lax canopy, it is ideally suited for small farm agroforestry systems. Its natural distribution throughout India is indicative of its adaptation to diverse agro-climatic conditions. Farmers are also familiar with jatropha as a hardy fence plant that can survive with very little inputs or management. These advantages notwithstanding, it should not be assumed that jatropha can succeed under any condition.

Another advantage of jatropha is the properties of its oil for biodiesel production. Its low free fatty acid content, almost similar to edible oils, makes it ideal for transesterification and the oil to biodiesel conversion ratio is higher than in other non-edible oils. A higher ratio means better profit margins and hence jatropha oil is a more desirable raw material than others. Combustion studies have also shown jatropha biodiesel to be superior to others as its emissions are less polluting.

Pongamia has its own merits as a TBOS. Rural communities in India are familiar with this species because its oil has been used traditionally for lighting lamps in households. It grows well in most parts of the country. A particular merit of this species is its ability to withstand both waterlogging and extreme drought conditions. In dry areas, it is one of the few species remaining green during the summer season. Its ability to fix atmospheric nitrogen is another advantage of this species. Domestication work on this species to identify elite genotypes and standardise cultivation practices is still at an early stage. Because this species is native to India, most of the information related to its use as an economically important plant and contribution to scientific knowledge has to be generated locally.

Neem is a species that combines two environment-friendly themes of current interest in the form of biopesticide and biodiesel. It is a hardy species that not only suvives, but produces reasonable quantities of seed in all types of environments. Of particular significance is its ability to grow on saline soils. Its wood is of reasonably good quality timber that can be used for making farm implements and construction purposes. It can also combine well with many annual crops as well as small trees like jatropha in agroforestry systems.

PROSPECT OF AGROFORESTRY IN INDIA

Trees and forests were always considered as an integral part of the Indian culture. The ancient scriptures and historical records amply support this. The best of Indian culture was born in the forests. The Aryan civilization was started in our forests and our Rishis who evolved the Hindu religion, lived in forests in complete harmony with nature. The ashrams were the centers which harmonized agriculture and pasture with trees, animals and birds. It was widely believed that destruction of forests and cutting of trees created famine conditions, where as planting and maintaining trees were regarded as noble acts. In fact, so much has been written in our ancient literature that individuals on their own agricultural fields were doing planting tree in ancient times.

Gradually, during recent periods because of increasing population and huge gap between demand and supply, forests were ruthlessly exploited to meet the increasing demand of fuel, fodder and timber. To overcome this huge burden upon our existing forests, some alternative steps have to be taken to meet the increasing demand of forest produce i.e. production of such items have to be carried outside the forest areas as well. Hence, in the light of ever increasing demand, concept of multiple use of land with multipurpose tree species has become immensely important. In this context, agro forestry, which is a form of multiple land/use system, should be adopted and encouraged.

The reasons for higher production under agro forestry system include:

- Greater efficiency of tree species for photosynthesis.
- Improved soil structure and fertility with increasing effects on crop yield.
- Reduce losses from soil erosion and more closed cycling of organic matter and nutrients.
- Creating better micro climatic conditions for the growth of agricultural crops.

AGROFORESTRY: CONCEPT AND DEFINITION

Agro forestry is not a new system or concept. The practice is very old, but the term is definitely new. Agro forestry means practice of agriculture and forestry on the same piece of land. Bene et al. (1977) defined agro forestry as a sustainable management system for land that increases overall production, combines agricultural crops and animals simultaneously. Nair (1979) defines agro forestry as a land use system that integrates trees, crops and animals in a way that is scientifically sound, ecologically desirable, practically feasible and socially acceptable to the farmers. Another widely used definition given by the International Center for Research in Agro forestry (ICRAF) Nairobi, Kenya, that, "agro forestry is a collective name for all land use systems and practices where woody perennials are deliberately grown on the same land management unit as agricultural crops or animals in some form of spatial arrangement or temporal sequence" (Nair, 1983).

Beneficial Effect of Agroforestry

Higher yields of crops have been observed in forest-influenced soils than in ordinary soils. In the Tarai area of Uttar Pradesh, Taungya cultivators harvested higher yields of crops such as maize, wheat, pulses etc. without fertilizer. Approximately, 20 per cent higher yields of grains and wood have been reported in agro forestry areas of Haryana and western Uttar Pradesh than from pure agriculture (Dwivedi and Sharma, 1989).

Experiments conducted at IGFRI, Jhansi indicate that the total yield of fodder is more when fodder grasses are grown with fodder trees than pure fodder grass cultivation. Leucaena leucocephala intercropped with agricultural crops and fodder grasses increase the total yield of food grains, fodder and fuel (Pathak, 1989).

Nitrogen fixing trees grown in the agro forestry systems are capable of fixing about 50 -100 Kg N/ha/year (Tewari, 1995). Experience in Punjab, Haryana, Uttar Pradesh, Gujarat and some parts of the southern states indicate that a tree and agriculture crop production system is more productive. The total production and value of fuel, fodder and small timber in degraded lands are reported to be many times more than the coarse grains usually produced on them (Gupta and Mohan, 1982).

Sanchez (1987) stated that, " appropriate agro forestry systems improve soils physical properties, maintain soil organic matter and promote nutrient cycling". Nitrogen fixing trees are mentioned as one of the most promising component of agro forestry system. The leaf litter after decomposition forms humus, releases nutrients and improves various soil properties, it also reduces the fertilizer needs.

Growing of trees and fodder crops (including fodder trees) is more economical, particularly on marginal lands. Observations taken in hot arid and semi-arid areas of Rajasthan indicate that marginal lands are incapable of sustaining stable and dynamic cultivation of agricultural crops. Silvipasture consisting of growing trees such as Prosopis, Albizia, Zizyphus and Acacia species may provide many times more returns per unit of land than agriculture under such conditions (Gupta and Mohan, 1982). Eucalyptus in agro forestry has been found to be more profitable than pure agriculture in Haryana. Populus deltoides increases the farm return by 50 per cent in Tarai region of Uttar Pradesh (Chaturvedi, 1981).

Agroforestry in Uttar Pradesh

After creation of Uttaranchal state in the year 2000, the tree cover in Uttar Pradesh has reduced to only 4.46 per cent where as, the State Forest Policy 1998 envisaged that one third of the total geographical area should come under forest/tree cover. Hence, agro forestry is now the only option to increase the desired tree cover of 33 per cent. In Uttar Pradesh, practices of agro forestry vary considerably according to the agro climatic zones, socioeconomic conditions and site-specific tree species.

The benefits of agro forestry is better understood by the farmers in western region of the state, this may be attributed, to the assured market of agroforestry produce because of flourishing wood based industries in the region. Eucalyptus and Poplar are preferred species in the western region, whereas, Shisham and Teak is preferred species in eastern region. Fruit trees also have considerable share of agro forestry particularly in western part of the state.

Hence, to workout suitable agro forestry models with preferred timber, fodder, fuel and fruit tree species for different agro climatic zones of U.P. a state level workshop was conducted by the research circle of U.P. forest department at Kanpur. The aim of the workshop was to bring together different workers including forest officers, scientists, subject matter specialists and NGOs working in the field of agro forestry at different places.

Suitable recommendations for tree-crop combinations in four different agro climatic zones of the state viz. tarai region, western plain gangetic region, eastern plain gangetic region and vindhya and bundelkhand region have been made. The outcome of the workshop held on dated 5-6th May 2001 was published and circulated under Lab to Land leaflet series entitled "Significance and use of Agro forestry System".

The tangible and intangible benefits of agro forestry as suggested in the leaflet are mentioned below:

- To meet the demand of fuel, fodder and timber for the increasing population.
- To reduce the biotic pressure on existing forests.
- To obtain maximum output in terms of yield from the same piece of land.
- To develop wasteland/degraded lands by planting suitable tree species with agricultural crops.
- To reduce the environmental pollution by planting tree species.
- To reduce soil erosion.
- To increase the soil fertility by planting nitrogen fixing tree species.
- To create availability of raw material for wood based industries.
- To create opportunity of employment to local people and to increase the return in terms of money by increased crop production.

AGROFORESTRY PROMOTION

The National Agriculture Policy, (2000) clearly states, "Agriculture has become a relatively unrewarding profession due to generally unfavorable price regime and low value addition, causing abandoning of farming and increasing migration from rural areas." Hence the Policy stresses, "Farmers will be encouraged to take up farm/agro-forestry for higher income generation by evolving technology, extension and credit support packages and removing constraints to development of agro forestry". Rural people have been practicing tree planting in their farms and homesteads to meet household requirements of fuel, poles, timber and medicinal plants. With the advent of social forestry, diversification in agriculture was encouraged to generate high income and minimize risks in cropping enterprises.

Planning Commission, GOI, 2001 for promoting agro forestry, has recommended the following:

- Rather than having a uniform strategy for the whole country, commercial agro forestry should be adopted in irrigated districts of the country.
- A separate strategy should be developed for rain fed areas for environmental security, sustainable agriculture (production and economy) and food accessibility.

- Suitable species for commercial agro forestry may include Acacia nilotica, Bamboo species, Casuarina equisetifolia, Eucalyptus species, Populus deltoides and Prosopis cineraria for different climatic, edaphic and agricultural conditions.
- Specific institutes have been identified for tree improvement and development of clones of specified species.
- Corporate private sector may be encouraged to take up research and development in tree improvement, development of better clones and micro and macro propagation of quality planting material.
- About 100 NGOs may be identified to carry out clonal propagation of seedlings for distribution to farmers at appropriate price and carrying out extension work. Extension activities should include organizing farmers, providing them training in planting techniques, protection measures and other silvicultural operations.
- Technological development to diversify usage of agro forestry species will help to ensure a ready market; for example bamboo is getting rediscovered as a potential raw-material for the development of bamboo composites suitable for use in place of wood and wood composites.
- Bamboo technology mission should be started keeping in view the impending gregarious flowering, followed by mass mortality of bamboo, forest fire famine and insurgency. Circumstances warrant formulation of emergency plans for harvesting and processing of bamboo prior to their flowering.
- As more and more farmers are taking up agro forestry, export - import policies should be modified to encourage agro forestry product marketing.
- A system of market regulation to be put in place, including a mechanism of periodic review in order to protect the interest of both producer and consumer of agro forestry produce.
- A suitable market information system needs to be introduced to inform farmers about major buyers, prevailing prices trends, procedure etc.
- All existing laws executive orders relating to tree felling transport, processing and sale should be amended to facilitate agro forestry.
- Commercial agro forestry may be planned in irrigated districts covering 10 m ha. On annual basis, one million ha should be brought under multipurpose tree species identified by the Task Force. The scheme of NABARD for farm/agro forestry should be expanded and investment of ₹. 100 crores per year should be ensured.

- It is proposed to cover 18 million ha of rain fed areas on watershed basis under agro forestry for conservation of soil and water and plantation of hardy species such as Eucalyptus, Bamboo and Babul. On annual basis 1.8 million ha is proposed for afforestation under various schemes of Rural Development, NAEB and 'food for work' scheme. An investment of ₹. 2700 crores will be required on yearly basis.
- Major states may establish Agro forestry Cooperative Federation for increasing bargaining powers of farmers in marketing of agro forestry products.
- Wood based industries should continue supply of quality planting material to farmers and ensure suitable buy-back arrangement.

PROFILES OF IMPORTANT TBOS

Considered to be a species of high potential for oilseed production, jatropha is an ideal species for wasteland development programmes. It is a small tree of about 6.0 m height belonging to family Euphorbiaceae. Its natural distribution is in Mexico and the Amazon region. It has a short trunk with thick branches spreading into a crown of dark green leaves. The bark is pale brown and the leaves are attached to long petioles. Native to humid zones under arid and semi-arid conditions, jatropha thrives under a wide range of soil and climatic conditions. It grows under annual average rainfall of 480 mm to 2400 mm and its daily average temperature range is 20–28°C. It tolerates extreme temperature conditions as well. It is drought tolerant and can withstand slight frost.

Jatropha can grow on almost any site, ranging from gravel, sandy to clayey soils. But its growth form is stunted in highly eroded soils of low fertility and in alkaline soils. It tolerates drought by shedding the leaves. This, however, results in decreased growth. It grows at altitudes ranging from sea level to 1000 m. Jatropha comes into flowering during September to December and fruits mature 2-4 months after flowering. However, in irrigated or high rainfall areas, a main harvest in October and a second harvest in March/April are possible.

Jatropha can be established with seeds or cuttings. Seeds can be sown directly at the onset of the rainy season or seedlings can be raised in polybags and transplanted. Branch cuttings grow vigorously when used as planting material. Cuttings of 45-100 cm in length and 3-4 cm in thickness taken from the base of the stem are the best for vegetative propagation. Plants raised from cuttings start bearing within one year whereas those propagated from seeds bear in 3-4 years. Canopy development by periodic pruning in the first three years is necessary to ensure high yields subsequently. Flowering occurs on one-year old branches. Therefore, after the tree enters the seed production phase, selective pruning must be done after harvesting the crop so that the new flush is mature enough to bear seeds during the next rainy season. Apart from the use of seed oil for biodiesel production, parts of jatropha have several other uses. The latex, oil, twigs, wood and leaves have medicinal value. Leaves are used as a fumigant for bed bugs. Different parts of jatropha are also used as pesticides. There are reports of leaves being used as a feed for silkworms. A dye extracted from leaves and tender stems is used for colouring cloth and fishing nets. Tender foliage and oil cake of jatropha can be used as organic manure.

Pongamia Pinnata

Pongamia pinnata, belonging to family Leguminosae, was earlier known as *Derris indica* and *Pongamia glabra*. It is a nitrogen fixing tree that produces seeds containing 25–30 per cent oil. It is often planted as an ornamental and shade tree. It is native to India and is receiving a widespread attention at present as a TBOS. Pongamia is a medium-sized tree that generally attains a height of about 8.0 m and a trunk diameter of more than 50 cm. The bark is thin, gray to grayish-brown, and yellow on the inside. The alternate, compound pinnate leaves consist of 5 or 7 leaflets which are arranged in 2 or 3 pairs, and a single terminal leaflet. Pods are elliptical, 3-6 cm long and 2-3 cm wide, thick walled, and usually contain a single seed. Seeds are 10-20 cm long, flat, oblong, and light brown in color.

Pongamia thrives in areas having an annual rainfall ranging from 500 to 2500 mm. In its natural habitat, the maximum temperature exceeds 38°C and the minimum can be as low as 1°C. Mature trees can withstand water logging and slight frost. This species grows in elevations of 1200 m, but in the Himalayan foothills it is not found above 600 m. It can grow on most soil types ranging from stony to sandy to clayey, including dry sands and saline soils. The natural distribution of this species is along coasts and river banks in India and Burma.

The average seed oil content is about 25 per cent, but higher percentages are claimed nowadays because of its use for making biodiesel. The oil is thick and yellow-orange to brown in colour. Traditionally, besides the cooking and lighting uses in rural areas, it was used as a lubricant, water-paint binder, pesticide, and in soap making and tanning industries. The oil is known to have value in folk medicine for the treatment of rheumatism as well as human and animal skin diseases.

Wood is beautifully grained and medium to coarse textured, but is not durable as it is susceptible to insect attack and tends to split when sawn. Therefore, it is used as fuelwood and cheap timber. The leaves are not readily eaten by animals, but it has some fodder value in dry areas. Leaves are also used as insect repellent in stored grains. The oilcake has use as poultry feed and also as manure with nematicidal value.

Azadirachta Indica

Azadirachta indica (neem) of family Meliaceae is native to dry forests of South and Southeast Asia. It has been receiving wide publicity because of the pesticidal properties of azadirachtin and other such constituents in its seeds. Neem can also fit into agroforestry systems as it is a multipurpose tree. It is a medium to large tree of about 20 m height. It can grow on a wide range of soils and climatic conditions. Neem grows on all types of soils, but prefers deep clay soils.

It can survive on acidic soils as well as those having pH of 10. The temperature range in its native range is between 15-45°C. Best growth of neem is found in areas receiving 750-1000 mm annual rainfall. In India, it grows in tropical dry areas up to an elevation of 1200 m. Neem seed has a very low period of viability and has to be sown within a month after harvest. Stump planting is the best method for neem establishment.

Neem leaves also contain the compounds found in the seed, but their concentration is very low. Therefore, the tree is not totally free of pests. Neem trees usually flower in April and the fruits mature 2-3 months later. The kernel contains about 30-40 per cent oil, which has the same range of uses as that from other TBOS. The seed cakes can also be used as manure. Recent advances in processing technologies can remove the bitter constituents in oil and seedcake and thereby widen their uses. The wood is strong and can be used for furniture, implements and construction. Seed, leaves, bark and fruit pulp have medicinal properties.

Madhuca Species

Madhuca longifolia and *Madhuca indica* belong to family Sapotaceae and are native to India. Although their seed oil is considered to be non-edible, they are used for cooking in tribal areas. The trees are generally decidous, light demanding and reach a height of up to 18 m and a diameter at breast height of 80 cm. The bark is grey to black with vertical cracks. Leaves are oblong shaped with a pinkish tinge and have a wooly underneath. Flowers are pale yellow and the flesh of the fruits is juicy.

Fruits are 3-5 cm long and contain 1-4 seeds. Commonly found in dry tropical and sub-tropical regions, mahua is found in areas receiving annual rainfall of 550 to 1500 mm. The temperature extremes it can tolerate are 2-46°C. It can grow in elevations up to 1200 m. This species requires deep loamy or sandy loamy soil with good drainage. It also grows well on shallow, clay and calcareous soil.

Besides the use of oil for cooking, fresh leaves of mahua are eaten as a salad and its ripe fruits are used for fermenting liquor. During periods of fodder scarcity, the foliage is used as fodder. Seeds contain 20-50 per cent oil and the seed cake has use as manure. Mahua oil is also used by rural communities for protection against storage pests. Almost all the parts of mahua tree have medicinal properties. Its reddish brown hardwood is strong and durable, and is used as timber.

Calophyllum Inophyllum

Calophyllum inophyllum is a medium-sized tree that sometimes reaches more than 25 m tall and up to 150 cm in diameter. It is a species for coastal environment. It thrives in areas having an annual rainfall of 750-5000 mm. In its natural habitat, the temperature rang is 7-48°C. This species grows at elevations up to 200 m. Calopyllum is sensitive to frost and fire. It prefers deep soils near the coast and can thrive even on pure sand. Its leaves are elliptical, thick, smooth and polished. Fruits are spherical to ovoid, greyish-green in colour and smooth skinned. Seeds have large cotyledons and the oil content of dry kernels is about 70 per cent.

The fruit is edible; usually it is pickled but care must be taken as it contains toxins. The strong and durable wood is used for boat building, railway sleepers, implements, instruments and construction. Oil has medicinal uses in treating rheumatism, ulcers and skin diseases. The latex is used internally against diarrhoea and externally against skin and eye diseases. Leaves, flowers and seeds are sometimes also used in local medicine.

PRODUCTION POTENTIAL OF KEY TBOS

Promotional efforts make out jatropha to be a wonder crop. Not all the information circulated to highlight its merits is based on hard evidence or actual field observations. Common beliefs, distorted claims and wishful thinking are all part of the information dissemination that jatropha is presently going through. Jatropha can be a success only if its potential is realised as an oilseed crop. The experiences of many who make unrealistic claims are based on observations of jatropha as a plant and not in a cropping situation.

Highlighting the advantages of pongamia over jatropha, its proponents present a strong case for this species to be the mainstay for TBO production in the country. Apart from its merit as an indigenous nitrogen-fixing tree, it has certain other favourable traits such as tolerance to waterlogging. However, as in the case of jatropha, not all what is presented can be taken at face value as some of the claims are yet to be substantiated. A few years back neem received the same type of publicity that jatropha is receiving at present. The validity of the claims pertaining to these species is discussed below.

Hardy Species

It is common knowledge that jatropha is a hardy species that can survive under extremely harsh environmental conditions with little or no external inputs and management. Similarly, pongamia survives under exceedingly dry conditions and often appears luxuriant even under the most stressful condition. But mere survival is not what is expected of a crop when it is cultivated for oilseed production.

In most farms where it is grown for livefencing, jatropha is planted as a single row. Pongamia and neem grow as scattered single trees or in small patches that probably are better in soil conditions than the neighbouring areas.

The presence of trees in a very low population density or growing wild do not necessarily indicate their ability to survive anywhere. Moreover, the hardiness of a species can be ascertained only when it produces repeated high yields for which it is grown. Without evaluating the productivity of these species as plantation crops in diverse agro-climatic conditions for a few years, they cannot be concluded to be hardy species.

Planting Material

The rapid rise in the interest to grow jatropha and pongamia has created a huge demand for seed, seedlings and vegetative propagules. A natural extension of this demand is the emergence of planting material suppliers who claim to have elite varieties of these species. It is not known whether these so-called varieties have undergone performance testing anywhere. The difference in performance observed among three genotypes of jatropha (BAIF, unpublished).

Although the local genotype outperformed the others, its superiority can be confirmed only if this performance is repeated in many locations over a period of time. Seed oil content is another parameter about which inflated claims are made. Generally, oil content of jatropha kernel is around 25 per cent, but some researchers have reported twice as much oil in seed collections from natural stands. It is too early to say whether such high oil contents can be achieved in large-scale cultivation. Moreover, oil content by itself is not of value unless it is associated with high yield of seed.

Another issue for consideration is the use of vegetative propagules as planting material. There are nurseries that sell propagules at exorbitant rates with the promise that they will come into seed production early. It is an established fact that propagules from cuttings or grafts of any species start bearing early and show uniform growth.

Vegetative propagation has been very successful in plantation crops like tea and rubber, but the recommended clones of those species have undergone testing for long periods under biotic and abiotic stresses. The suitability of vegetative propagules for establishing plantations of TBOS is untested. Until studies are conducted on this aspect, seedlings that have the tap root system may be a safer option for wastelands than propagules with adventitious roots.

Seed and Oil Yield

The most appealing aspect of jatropha for unsuspecting growers is the claim of promoters that a seed yield can be had in the first year itself and eventually when the trees mature in 3-4 years, the yield can be as high as 8.0-10 tons per ha per year. These are highly optimistic projections. Although it is a perennial, jatropha starts flowering in the first year itself and may produce some seeds. Harvesting these seeds is undesirable because the priority for the first three years of establishment of a commercial plantation is to develop the tree canopy.

This is achieved by periodic pruning and thinning of branches. During this juvenile phase, flowers that appear should be removed at the bud stage so that vegetative growth continues unaffected. Thus, the information circulated with regard to the time of first commercial harvest and the seed yield at maturity is not unrealistic.

Similarly, yield and age at first seed harvest projected for pongamia are also disputable. Seed production of pongamia will be small for at least six years. There are claims that each pongamia tree will yield a minimum of 20 kg and the per ha production will be more than 10 t. These are highly optimistic expectations and will require fertile land with irrigation and high levels of nutrient application, negating the argument for production of oilseeds on wastelands.

Seed oil content is another parameter about which misleading data is in circulation. There are some claims of more than 50 per cent oil and the economics of production are calculated accordingly. It does not appear that there are varieties or genotypes that can yield such high levels of seed oil when cultivated commercially. Under certain conditions or certain genotypes may be having such levels of seed oil content, but their seed yield may be low. Therefore, seed oil produced per unit area of land, which combines both seed yield and seed oil content, is a more useful indicator of productivity of TBOS.

Inputs

The belief that jatropha, pongamia and other TBOS grow with little or no inputs is also a popular perception and does not apply to commercial cultivation. Repeated harvest of seed in substantial quantities will drain the soil as well as the plant of nutrients and moisture. Unless these are replenished through fertilizer or manure application, seed production cannot continue. Like other crops, these species will also require water to grow and be productive. The seedlings will require irrigation during the first two years in the field.

Thereafter, once the root system grows deep and establishes well, it will be able to make use of water held in lower layers of soil. If the water holding capacity of the soil is low, regular irrigation will be necessary for regular seed harvests to be obtained. If the goal is to produce seed on a commercial scale, input supply has to match the crop requirements. Such requirements are usually much higher for trees grown for the purpose of production than those serving the function of live fencing.

Multipurpose Use

All the TBOS are multipurpose in their utility, making them what is desired for agroforestry systems. However, caution is necessary in assessing whether all the uses will be realised at the same time. For example, jatropha has several other uses besides the seed oil. Almost all its parts have medicinal properties and are used in native medicinal preparations. Extracts of its bark can be made into a wax. Tannin or dye can be extracted from leaf and root of jatropha.

These characteristics are projected as additional benefits from the multipurpose jatropha tree. In reality, if plant parts are harvested for other purposes, seed yield will be adversely affected. Therefore, commercial cultivation should focus on the seed; periodically pruned branches and fallen leaf litter should be considered as by-products from the jatropha plant. In addition, others like oilseed cake and glycerol will be obtained as by-products of seed processing.

Productive Lifespan

The lifespan of jatropha, according to most literature, is 40-50 years and that of pongamia is more than 80 years. It is possible that trees of these species can survive this long when regular harvests are not taken from them. Growing them as commercial plantation crops will result in regular harvest of seed. In addition, there will be growth stimulation in the form of periodic pruning and application of fertilisers or manures.

All these practices will make the tree weaker. Experience with plantation crops like tea and rubber show that even if the trees survive, beyond a particular stage their productivity will decline to such an extent that it will not be economical to retain them. Therefore, it is safer to assume the productive life for jatropha and pongamia to be about 20-30 years.

PRODUCTION POTENTIAL

Realistic Goals

The manner in which TBO production is promoted at present with exaggerated claims is fraught with the possibility of growers losing interest if the very high expectations are not realised. This has happened in other crops in the past. It is not that these crops failed altogether, but their potential was blown out of proportion and farmers were unhappy with what they eventually realised. Even jatropha went through such an experience when farmers discontinued its commercial cultivation in Maharashtra a few years ago. However, the current situation and the purpose for which it is promoted are totally different, so the chances of its success are greater this time around.

Information to set realistic production targets for TBOS is scanty, but approximate figures are possible for jatropha. Intensive jatropha production on good agricultural land with drip irrigation, high levels of fertilisers or manures and adoption of management practices such as regular pruning may yield 5.0 t per ha or more. However, a farmer getting into jatropha production cannot start off with an expectation of 5.0 t per ha or more. Plantations of this intensive category have been established in recent years, but they are mostly still in their juvenile stage and actual production data is unavailable at present.

The other extreme is a below-average land that is managed less intensively to yield about 1.0 t per ha. This level of yield does not hold out much promise for a prospective grower to enter into TBO production. In between these two extremes is the production under average conditions of soil fertility and management. With presently available planting material and technology package, a modest target that has a high probability of attainment is 2.0 t per ha. Therefore, the short to medium-term oilseed production strategy has to target a potential yield of 2.0 t per ha on land of average to marginal productivity.

Although popular literature claims a seed yield advantage of more than 50 per cent for pongamia over jatropha, in the absence of hard evidence, the yield levels of both species have to be assumed to be almost the same. In terms of suitability for agroforestry with annual crops, jatropha with its sparse canopy appears to be more suited. It also responds well to training of the canopy by trimming and pruning (Banerjee, 1989). If the system is a combination of perennials, pongamia is likely to fit in better.

TBOS-Based Agroforestry Systems

The success of any agroforestry system depends to a great extent on the compatibility among the component species. In situations where jatropha or pongamia grows on farmland, a negative effect on the growth of crops growing nearby can be seen. This may not be apparent during the first five years, but becomes pronounced with time. This may be partly overcome in mature trees by canopy management practices like pruning. Studies conducted at BAIF to examine the suitability of jatropha for agroforestry purposes showed that the yield reduction on horse gram and finger millet was negligible during the first four years. In fact, the annual crops seemed to benefit from the sheltering effect of jatropha when there were strong dry winds.

Agroforestry systems can be in the form of (a) scattered trees, (b) rows or strips of trees and (c) trees in border rows or strips. The ratio of trees to crops in the system can be determined by adjusting the spacing between the trees as well as spacing between rows or strips. The present practice of a single row of jatropha livefencing can be extended to a strip of 3.0-5.0 m. This strip can accommodate 2-3 rows of jatropha or a single row of pongamia. This would effectively mean devoting up to 20 per cent of land to the tree component. This arrangement will take up relatively less land for the tree species while the protection function of the livefence will be strengthened.

The outcome of this arrangement would be a loss of about 25 per cent in the yield of the crops in the middle of the field while the fence produces oilseed. Farmers may not opt for this system if the land is fertile because the loss in crop production is unlikely to be compensated by the TBOS in the fence. On the other hand, it can be a favourable alternative for lands where crop yields are low and less reliable.

The other two systems, scattered trees and strips within the cropping area, are also unsuitable for fertile lands because the trees can reduce the crop yields. On such lands, fruit trees may be acceptable as the produce is of high value. But the returns from TBOS are unattractive for farmers to forego existing crops that are proven and more remunerative. Depending upon the fertility, a suitable tree:crop ratio can be decided upon and the tree spacing can be arrived at. If the tree proportion in the system is assumed to be 50 per cent, the jatropha yield, as discussed earlier for average production conditions, will be about 1.0 t per ha.

This translates to an income of ₹. 10,000 per ha (2000 kg per ha at the rate of ₹. 5 per kg of seed), which is not appealing enough to attract growers in large numbers. But it is also a fact that present returns from vast areas of wastelands in the country are lower than this amount. Most of these lands belong to small marginal farmers in dry areas who routinely take up cultivation of some crop during the rainy season. Jatropha and pongamia are suitable options for them. There are others who raise crops on a part of their land while the less fertile part usually remains barren. Such land can be brought under TBOS as it requires very little management after establishment and will not be a constraint on farmer's time and resources.

Concluding Remarks

Presently, tree-borne oilseed species, jatropha in particular, find themselves in a rare situation where several factors are in their favour. Those of significance are: the record high prices of petroleum products and the rapid depletion of their global reserves; worsening environmental problems and the emerging stringency in the emission standards of fast-growing economies; the expanse of degraded barren land in India that has to be brought under vegetation for the future common good; an opportunity to earn carbon credits through Clean Development Mechanism while creating rural employment and a non-conventional energy system. Opportunities such as this with so many factors in favour do not come up that often.

Considering the prevailing favourable situation, the biodiesel programme based on TBOS should not be allowed to falter at this stage. A lot is heard about government programmes for their promotion, but there appears to be a lull at implementation level. The non-government efforts tend to be driven by overly optimistic or deliberately exaggerated production levels. TBOS can succeed if promoted through a systematic programme with realistic targets.

COMMON CLASSIFICATION OF AGRO- FORESTRY

AGRI-SILVICULTURE

This system emphasizes rising of trees and cultivation of food crops and fodder crops in the available space between the trees. In the plantation of Leucaena. Field crops such as pigeon pea + back gram are grown successfully with least reduction in fodder yield of trees component. The wood yield of eucalyptus was found to increases in association with the intercrops of cassava + groundnut in Kerala in sloping laterite soils. Acacia Nilotica is raised on the bunds of rice fields and has high tolerance to water logging. It grows rapidly getting benefits from irrigation and fertilizer given to rice crop.

Silvi – Pasture

In the silvi- pastoral system, improved pasture species are introduced with tree species. In this system, grass or grass – legume mixture is grown along with the woody perennial simultaneously of the same unit of land. This is the best management for the wastelands, with marginal fertility. Multi- layered vegetation covers are very effective in controlling runoff and soil loss from erosion- prone areas. This system not only reduces surface runoff but also facilate infiltration and storage of rainwater in the soil profile. Grass Species: It should be palatable, high yielding and should be tolerate to shade and with good competing ability. Rotational grazing is effective in increasing the grazing potential and efficiency. Some suitable grasses are: Pennisetum pedicellatu, Cenchrus ciliaris, c. sehima nervosum, chrysopton fulvus, and Dichanthium annulatum.

Agri–Horti System

Agricultural crops are normally grown in the interspaced of fruit trees planted at a spacing of 5–7 m apart. The fruit trees are managed for 30–35 years and they give regular income. In addition, they also provide fuel wood and fodder. Intercultivated agriculture crops proOvide seasonal returns. Fruit tress such as guava, custard apple, Ber phalsa, Jamun, bael, wood–apple and pomegranate are found promising under dryland situation. On marginal and wastelands, ber anona and citrus shown good promise. Millets, pulses or oil seeds can be grown as intercrops under fruit in the semi- Arid Tropics.

Agri–Silvipasture

It is the combination of agri–silviculture and silvi–pastoral system. Under this system farmers in dryland grown field crops and forest trees together up to a particular stage but in a later stage, the grasses are raised in place of field crops in the vacant space between the forest trees. Thus, there resource–poor farmers are assured if wood, agricultural products and grasses.

Agri–Horti–Silvivulture

In this system, fruit trees are grown along with crops and multipurpose tree species (MPTs). Agri–Horti–Siliviculture is highly dicers in vegetation with highest productivity up to 25.8 t/ ha/ year.

Homestead Agro Forestry

Forestry generally plant tree in and around their habitations, court yard, threshing floor and in the filed. These home gardens are aimed to satisfy the family needs of fuel, fodder and small timbers. They also fulfill the needs of fruits, vegetable, spices etc. The system of home gardens is, more prevalent in high rainfall area of Kerala and Tamil Nadu. In these regions, coconut and cash nut are the main crops. Several of animal are supported in most of the home gardens. The foliage from the is used as fodder.

3

Tropical Forests

INTRODUCTION

Primary tropical forest is not easy to find, although much has been so labeled by certain writers. In his excellent treatise on the world's tropical rain forests, Richards adds a most revealing footnote. In discussing what, in the original text, he characterises as "Primary Mixed Forest" in Nigeria, he remarks that further work in the area "makes it seem likely that this forest has suffered disturbance in the past and is old secondary rather than truly primary." He adds: "The same is probably true for nearly all so-called virgin or primary forest in Nigeria and perhaps the whole of West Africa." Undisturbed primary forest is usually thought to be of considerable extent in the Amcrican tropics; but it is by no means so widespread as is generally supposed.

The *Tropical Rain*

Forest is perhaps our least understood vegetation type. Many forest areas in the tropics classed as "rain forest" actually are *Light Tropical Forest.* A further complication is the presence of large forest tracts on mountain slopes in the tropics which, because of the terrain and wind pattern, are well watered. These are *Tropical Montane Rain Forests;* on the middle slopes where the precipitation is usually highest, they have a composition considerably different from that of the rain forests of the lowlands. Where they are adjacent, both lowland rain forest and light forest pass into the montane rain-forest type.

However, if we take a rather narrow zone across the equator, including those areas in which rain forest would be expected to develop, we discover that it is advisable to particularise our information. It is immediately obvious that there is only a narrow strip along the equator in which rain can be expected to fall at almost all times of the year. Here seasonal maxima may occur, but only exceptionally does any long period pass without rain. This relatively narrow equatorial band is the locale of the true rain forest, although local conditions in some places produce equivalent conditions that enlarge the occurrence of this forest type. Furthermore, some of the areas commonly mapped as rain forest actually contain considerable tracts of light forest.

Wet and dry seasons begin to be evident at only a few degrees of latitude north and south of the equator. The total annual rainfall may be very great, but, on the whole, true rain forest does not exist where the dry seasons last longer than a few weeks. However, in such areas along the main river valleys and even along the smaller streams, the residual soil moisture is such that what might reasonably still be classed as rain forest does occur.

This readily explains why observers traveling only along streams have in the past often expanded the supposed general distribution of this type of forest far beyond its actual limits. Glimpses of forest were seen in the distance and it was supposed that it was of the same general composition as that near at hand. Actually this riparian "rain forest," intercalated with light forest, is comparable to the fingers of gallery forest that follow the streams in savanna regions.

Light forest is found in those regions near the equator where there are two yearly maxima of rain, with one of the dry seasons of about three months in length. With a shortening of the longer dry season, this forest type approaches the rain forest in character.

However, on well-drained to overdrained terrain, the amount of moisture retained-the "effective precipitation" — may be low enough to produce characteristic light forest relatively close to the equator (and, in fact, in some places actually athwart it). Where the rainfall pattern is such that there is only one rainy period and a correspondingly long dry period, light forest persists until the conditions are reached where natural savanna (or in some instances thorn forest and scrub) occurs.

The light forest is obviously not a single forest "type." It may be classified on a local basis, but in broad pattern it varies from a "wet" phase to a "dry" phase through myriad nuances of vegetational association. In discussing both the rain forest and what actually is the "wetter" phase of light forest, Richards indicates something of its nature. There are exceedingly complex ("mixed") types, with no species really dominant in the stands and there are also extensive forests with single dominant species.

He likens the "mixed" tropical forest to the mixed mesophytic forest of eastern North America, which develops only in areas with abundantly varied micro-environments. I suggest here that the "mixed" forest of the tropics may also be a type that develops primarily in areas with a series of varied micro-environments, as yet essentially unstudied and therefore not understood. The forests near major streams, where they are subject to periodic, or occasional, inundation — areas often many miles in width — are, of course, "disturbed forests" and therefore likely to be complex, with a characteristic mixed community and a considerable amount of "jungle" undergrowth. In the light forest, unless disturbed in some manner, the canopy is usually sufficiently close to prevent herbaceous and shrubby forms from receiving light enough to form more than a hint of jungle-like undergrowth.

On the whole, agriculture in the tropics is not successful in those areas characterised by typical rain forest, nor is life easy there. Hence, in his constant search for habitable places, man long ago entered the light forest, cleared the land and set about his routine chores of cultivating it. Great areas in India, once covered with "monsoon forest" (a typical form of light forest produced by the strong periodicity of rainfall), are now given over to intensive agriculture. The more primitive pastoral agriculture of central Africa has also eliminated large stretches of light forest, but here savanna has resulted, so that it is today most difficult to distinguish between natural savanna and that which has been artificially produced by fire and overgrazing.

Great as are the forest resources of the tropics, we have as yet made far too little intelligent use of them. At present, the world's primary demand is for the coniferous softwoods. Numerous especially valuable timbers are today being cut from tropical forests, but the great majority of the species are ignored. Remote from the daily experience of most of us, the spoilage of the remaining areas of tropical rain forest and light forest through improper practices is not generally comprehended.

Under our present economy many of the species of these forests are considered "worthless" and their normal habitats are sometimes almost wantonly destroyed. These "worthless" forms constitute at least a vast reservoir of raw cellulose; their other uses may be legion. A great natural resource rapidly being wasted away through ignorance.

Variability and Kind

The origin of forest types, those times in the history of the world when the land surfaces were elevated, when mountain building was at a maximum and when climatic disturbances were widespread marked important periods in the development of new vegetation types. In view of the changes that land surfaces and climates constantly undergo, a too rigid genetic system is a distinct disadvantage to any group of organisms. With genetic plasticity — variability -there is a greater chance of survival during any disorganisation of habitat. Variability, indeed, is the basis of survival.

From the standpoint of plant systematics the species is usually considered to be the working unit. Furthermore, it is traditionally supposed to exhibit but little variation. This concept, however, is largely based on superficial morphological grounds.

As we begin to study species in greater detail, examining thousands of individuals instead of only a few in each species, it becomes evident that even this so-called morphological constancy of the species has little basis in fact; we have not been scrutinizing our materials with sufficient precision. We also are learning that, with this morphological variability, there is an equally large amount of physiological variability, inherent in the genetics of the group; and, in the end, this inherited physiological variability is far more important than are differences in morphology where factors concerned with survival are involved.

Foresters have long known that, in reforestation or afforestation, it is best to use seed from sources not too far removed from the site of the new plantings. In western North America Douglas fir (*Pseudotsuga taxifolia*) grows from about latitude 55° in British Columbia to about latitude 170 in Mexico, from sea level in the Pacific Northwest to elevations of over 10,000 feet in the Rocky Mountains and from humid to subhumid habitats.

Attempts have been made to set up a series of "segregate species" from within the known morphological variability of this wide-ranging tree, but from the point of practical botanical taxonomy only the variety *glauca* is now recognised apart from the "typical" form. However, it has been found best not to move materials much more than a degree of latitude, or a climatically comparable distance in elevation, if the greatest survival and maximum growth of the planting is desired, though practical considerations usually make this impossible to achieve.

As a species defined by taxonomists, Douglas fir is widespread, but each of its local biotypes has a genetic norm that fits it to a particular set of ecological conditions. Even here we can discover considerable variability, related to the precise conditions of the local micro-environments. There also is a growing body of data that day length at certain periods of the year — which, of course, is correlated with latitude — plays a very important part in the effective competition and survival of these minor biotypes.

These and similar examples might be greatly expanded. The white pine (*Pinus strobus*) is found from eastern Canada through the Appalachian Mountains and also in the highlands of southern Mexico. When brought together at about latitude 40° N and if protected by appropriate handling under glass, the material from Mexico grows considerably faster than does that from the central Appalachians, as in turn this grows faster than do the Canadian plants. When planted in the open at this same latitude, the more nearly local material still grows faster than does the Canadian material, but the Mexican plants die with the first hard freeze. Data of the same sort are being accumulated for hardwoods that have wide distributions and also apply to adaptability to variant soils as well as to differences in climate-in brief, to the host of factors that influence the growth and persistence of plants.

The source of this genetic variability within species and the mechanics of its operation within species populations, which often enables the sum of their biotypes to have considerable geographical ranges and ecological tolerances, is a subject too extensive to be developed here. Nevertheless, what we can ascertain by relatively precise experimental methods with regard to what is happening among the species around us today, is indicative of what has been going on through the long span of time during which forests have covered the land.

Through such studies we come to an understanding of how it was that, in the past, groups of organisms could develop biotypic variations capable of venturing into new habitats and new climatic regions.

Thus we glean insight into the importance of the uplands and mountain ranges of the past. It was on the uplands and cooler mountain slopes of the frost-free regions that the ancestors of our modern forest trees evolved. On these same slopes, with their multitude of different micro-environments, variant forms better preadapted to conditions in temperate regions came into being, were further elaborated, could persist for long periods of time and so be available to spread poleward during those periods when the great cyclic changes in the topography of the continental masses and their climates afforded an opportunity.

Likewise, this same inherent tendency towards variation, characteristic of all life, gave rise to forms still better fitted to cooler or even to boreal climates. With a swing of the climatic pendulum in the other direction, the ancestral biotypes would necessarily have had to retreat, or, if unable to do so for some reason, become extinct, leaving behind the newly evolved forms. These derived forms are the backbone of the vegetation of our present temperate and cool-temperate forests. With further evolution, coupled with genetic isolation, these biotypic segregates could — and have — become the "boreal" and "extreme austral" species of the taxonomist.

There is a tendency of some people to think of the tropical lowlands, because of their generally equable conditions, as having had a persistent and stable series of forest communities over long periods of time.

This would seem to be a valid assumption on theoretical climatological grounds and also because considerable segments of the equatorial lowlands have been in a reasonably stable geographical position, at least since the Permian. However, there is evidence that the present forest types of the lowlands have been derived in large part from more upland types, which migrated into the lowlands during the Middle and Late Tertiary. As the writer has pointed out elsewhere, the forests of the central Amazonian basin appear to be primarily composed of elements derived from the surrounding uplands, perhaps at a time no earlier than the Late Pliocene or Early Pleistocene.

Obviously, based as it is on the consideration of a single area, this generalisation needs considerable study before it can be taken as being more widely applicable. On the whole, however, the forests of tropical lowlands do not contain what, according to generally accepted botanical criteria, are considered to be the more primitive living species of genera that have fairly broad altitudinal distributions. These are usually found on the piedmont and lower mountain slopes.

Proper utilisation of our forest resources demands that trees be cut down. What comes back to take their place-or what is brought in to replace a forest area that has been despoiled by improper cutting practices — is not a thing to be decided in a casual manner. Even with selective thinning, the micro-environments have been changed and with clear-cutting a totally new set of factors is introduced. Some few forest types regenerate satisfactorily; others are replaced — or might better be replaced — with different sets of materials. Whether the replacement is to be with the same biotypes, with somewhat different biotypes of the original species, or with entirely different species, depends on a considerable array of factors.

These factors are within the province of scientific silviculture, but they cannot be ascertained without a great deal of painstaking research.Biotypic variation is the fundamental mechanism leading to the evolution of the manifold kinds of organisms with which this world abounds and an understanding of it is indispensable in any consideration of the proper utilisation and regeneration of our forests, our most important replaceable natural resource.

VALUE OF TROPICAL FORESTS

All forests have both economic and ecological value, but tropical forests are especially important in global economy. These forests cover less than 6 percent of the Earth's land area, but they contain the vast majority of the world's plant and animal genetic resources. The diversity of life is astonishing. The original forests of Puerto Rico, for example, contain more than 500 species of trees in 70 botanical families. By comparison, temperate forests have relatively few. Such diversity is attributed to variations in elevation, climate, and soil, and to the lack of frost.

There is also diversity in other life forms: shrubs, herbs, epiphytes, mammals, birds, reptiles, amphibians, and insects. One study suggests that tropical rain forests may contain as many as 30 million different kinds of plants and animals, most of which are insects.

WOOD AND OTHER PRODUCTS

Tropical forests provide many valuable products including rubber, fruits and nuts, meat, rattan, medicinal herbs, floral greenery, lumber, firewood, and charcoal. Such forests are used by local people for subsistence hunting and fishing. They provide income and jobs for hundreds of millions of people in small, medium, and large industries.

Tropical forests are noted for their beautiful woods. Four important commercial woods are mahogany, teak, melina, and okoume. Honduras mahogany (Swietenia macrophylla), grows in the Americas from Mexico to Bolivia. A strong wood of medium density, mahogany is easy to work, is long lasting, and has good colour and grain. It is commonly used for furniture, molding, paneling, and trim. Because of its resistance to decay, it is a popular wood used in boats.

Teak (Tectona grandis) is native to India and Southeast Asia. Its wood has medium density, is strong, polishes well, and has a warm yellow-brown colour. Also prized for resistance to insects and rot, teak is commonly used in cabinets, trim, flooring, furniture, and boats figure. Melina (Gmelina arborca) grows naturally from India through Vietnam. Noted for fast growth, melina has light colored wood that is used mainly for pulp and particleboard, matches and carpentry. Okoume (Aucoumea klaineana) is native to Gabon an the Congo in west Africa. A large fast-growing tree, the wood has mod erately low density, good strength-to density ratio, and low shrinkage during drying. It is commonly use(for plywood, paneling, interior fur niture parts, and light construction.

Other Economic Values

Tropical forests are home for tribal hunter-gatherers whose way of life has been relatively unchanged for centuries. These people depend on the forests for their livelihood. More than 2.5 million people also live in areas adjacent to tropical forests. They rely on the forests for their water, fuelwood, and other resources and on its shrinking land base for their shifting agriculture. For urban dwellers, tropical forests provide water for domestic use and hydroelectric power. Their scenic beauty, educational value, and opportunities for outdoor recreation support tourist industries.

Many medicines and drugs come from plants found only in tropical rain forests. Some of the best known are quinine, an ancient drug used for malaria; curare, an anesthetic and muscle relaxant used in surgery; and rosy periwinkle, a treatment for Hodgkin's disease and leukemia. Research has identified other potential drugs that may have value as contraceptives or in treating a multitude of maladies such as arthritis, hepatitis, insect bites, fever, coughs, and colds. Many more may be found. In all, only a few thousand species have been evaluated for their medicinal value. In addition, many plants of tropical forests find uses in homes and gardens: ferns and palms, the hardy split-leaf philodendron, marantas, bromeliads, and orchids, to name just a few.

Environmental Benefits

Tropical forests do more than respond to local climatic conditions; they actually influence the climate. Through transpiration, the enormous number of plants found in rain forests return huge amounts of water to the atmosphere, increasing humidity and rainfall, and cooling the air for miles around. In addition, tropical forests replenish the air by utilising carbon dioxide and giving off oxygen. By fixing carbon they help maintain the atmospheric carbon dioxide levels low and counteract the global "greenhouse" effect.

Forests also moderate stream flow. Trees slow the onslaught of tropical downpours, use and store vast quantities of water, and help hold the soil in place. When trees are cleared, rainfall runs off more quickly, contributing to floods and erosion.

DEFORESTATION

Before the dawn of agriculture approximately 10,000 years ago, forests and open woodland covered about 15.3 billion acres (6.2 billion ha) of the globe. Over the centuries, however, about one-third of these natural forests has been destroyed. According to a 1982 study by FAO, about 27.9 million acres (11.3 million ha) of tropical forests are cut each year-an area about the size of the States of Ohio or Virginia. Between 1985 and 1990, an estimated 210 million acres (85 million ha) of tropical forests were cut or cleared. In India, Malaysia, and the Philippines, the best commercial forests are gone, and cutting is increasing in South America. If deforestation is not stopped soon, the world will lose most of its tropical forests in the next several decades.

Reasons for Deforestation

Several factors are responsible for deforestation in the Tropics: clearing for agriculture, fuelwood cutting, and harvesting of wood products. By far the most important of these is clearing for agriculture. In the Tropics, the age-old practice of shifting, sometimes called "slash-and-burn," agriculture has been used for centuries. In this primitive system, local people cut a small patch of forest to make way for subsistence farming. After a few years, soil fertility declines and people move on, usually to cut another patch of trees and begin another garden. In the abandoned garden plot, the degraded soil at first supports only weeds and shrubby trees. Later, soil fertility and trees return, but that may take decades. As population pressure increases, the fallow (rest) period between cycles of gardening is shortened, agricultural yields decrease, and the forest region is further degraded to small trees, brush, or eroded savanna.

Conversion to sedentary agriculture is an even greater threat to tropical forests. Vast areas that once supported tropical forests are now permanently occupied by subsistence farmers and ranchers and by commercial farmers who produce sugar, cocoa, palm oil, and other products.

In many tropical countries there is a critical shortage of firewood. For millions of rural poor, survival depends on finding enough wood to cook the evening meal. Every year more of the forest is destroyed, and the distance from home to the forest increases. Not only do people suffer by having to spend much of their time in the search for wood, but so does the land. Damage is greatest in dry tropical forests where firewood cutting converts forests to savannas and grasslands.

The global demand for tropical hardwoods, an $8-billion-a-year industry, also contributes to forest loss. Tropical forests are usually selectively logged rather than clear-cut. Selective logging leaves the forest cover intact but usually reduces its commercial value because the biggest and best trees are removed.

Selective logging also damages remaining trees and soil, increases the likelihood of fire, and degrades the habitat for wildlife species that require large, old trees-the ones usually cut. In addition, logging roads open up the forests to shifting cultivation and permanent settlement. In the past, logging was done primarily by primitive means-trees were cut with axes and logs were moved with animals such as oxen. Today the use of modern machinery—chain saws, tractors, and trucks -makes logging easier, faster, and potentially more destructive.

Endangered Wildlife

Forests are biological communities-complex associations of trees with other plants and animals that have evolved together over millions of years. Because of the worldwide loss of tropical forests, thousands of species of birds and animals are threatened with extinction. The list includes many unique and fascinating animals, among them the orangutan, mountain gorilla, manatee, jaguar, and Puerto Rican parrot. Although diverse and widely separated around the globe, these specles have one important thing in common. They, along with many other endangered species, rely on tropical forests for all or part of their habitat.

Orangutans (Pongo pygmaeus) are totally dependent on small and isolated patches of tropical forests remaining in Borneo and Sumatra, Indonesia. Orangutans spend most of their time in the forest canopy where they feed on leaves, figs and other fruit, bark, nuts, and insects. Large trees of the old-growth forests support woody vines that serve as aerial ladders, enabling the animals to move about, build their nests, and forage for food. When the old forests are cut, orangutans disappear.

The largest of all primates, the gorilla, is one of man's closest relatives in the animal kingdom. Too large and clumsy to move about in the forest canopy, the gorilla lives on the forest floor where it forages for a variety of plant materials. Loss of tropical forests in central and west Africa is a major reason for the decreasing numbers of mountain gorillas (Gorilla gorilla). Some habitat has been secured, but the future of this gentle giant is in grave danger as a result of habitat loss and poaching.

The jaguar (Leo onca), a resident of the Southwestern United States and Central and South America, is closely associated with forests. Its endangered status is the result of hunting and habitat loss. The Puerto Rican parrot (Amazona vittata), a medium-sized, green bird with blue wing feathers, once inhabited the entire island of Puerto Rico and the neighbouring islands of Mona and Culebra. Forest destruction is the principal reason for the decline of this species. Hunting also contributed. Today, only a few Puerto Rican parrots remain in the wild and their survival may depend on the success of a captive breeding programme.

In addition to species that reside in tropical forests year round, others depend on such forests for part of the year. Many species of migrant birds journey 1,000 miles or more between their summer breeding grounds in the north and their tropical wintering grounds. These birds are also threatened by tropical forest destruction.

THE PRACTICE OF FORESTRY

Forestry-loosely defined as the systematic management and use of forests and their natural resources for human benefit-has been practiced for centuries. Most often, forestry efforts have been initiated in response to indiscriminate timber cutting that denuded the land and caused erosion, floods, or a shortage of wood products.

Ancient Forestry Practices

In ancient Persia (now Iran), forest protection and nature conservation laws were in effect as early as 1,700 B.C. Two thousand years ago the Chinese practiced what they called "four sides" forestry-trees were planted on house side, village side, road side, and water side. More than 1,000 years ago, Javanese maharajahs brought in teak and began to cultivate it. In the African Tropics, agroforestry (growing of food crops in association with trees) has been practiced for hundreds of years. In the Yucatan Peninsula of southern Mexico, the ancient Mayas cultivated fruit and nut trees along with such staples as corn, beans, and squash. Bark, fibres, and resin were obtained from plants grown in fields, kitchen gardens, and orchards. Early in their civilization, the Mayas practiced slash-and-burn agriculture. As their population grew, they found more efficient methods of growing crops. They terraced hillsides, learned how to decrease the time between "rotations" of agricultural land with native forests, dug drainage channels and canals to move water to and from cultivated areas, and filled in swampland to plant crops.

The agricultural sophistication of the Mayas enabled their civilization to grow and flourish. What brought about their decline about A.D. 820 is not fully known, but some believe that as their society developed, the Mayas made unsustainable demands on their environment. Relatively little is known about tropical forestry before the mid1800's in most places. At that time, the European colonial empiresnotably the Dutch, English, and Spanish-brought modern forest management practices to Indonesia, India, Africa, and the Caribbean. Centers for forestry and forestry research were established, and more careful records were kept.

Sustainable Forestry

Modern forestry has its basis in 18th-century Germany. Like the Chinese and the Mayan forest practices, German forestry is essentially agricultural. Trees are managed as a crop. Two concepts are important: renewability and sustainability. Renewability means that trees can be replanted and seeded and harvested over and over again on the same tract of land in what are known as crop "rotations." Sustainability means that forest harvest can be sustained over the long term. How far into the future were foresters expected to plan? As long as there were vast acres of virgin (original) forests remaining, this question was somewhat academic. Today, however, sustainability is a vital issue in forestry. Most of the world's virgin forests are gone, and people must rely more and more on second- growth or managed forests. Perhaps we now face, as never before, the limits to long-term productivity.

In the German forest model, forestry is viewed as a continual process of harvest and regeneration. Harvest of wood products is a goal, but a forester's principal tasks are to assure long-term productivity. That is achieved by cutting the older, mature, and slow-growing timber to make way for a new crop of young, fast-growing trees.

Harvest-Regeneration Methods

Three examples of timber harvest-regeneration methods (silvicultural systems) illustrate how foresters manage stands to produce timber on a sustained basis.

Selection

Individual trees or small groups of trees are harvested as they become mature. Numerous small openings in the forest are created in which saplings or new seedlings can grow. The resulting forest has a continuous forest canopy and trees of all ages. Such systems favour slow-growing species that are shade tolerant.

Clearcutting

In clearcutting, an entire stand of trees is removed in one operation. From the forester's point of view, clearcutting is the easiest way to manage a forest-and the most economical. Regeneration may come from sprouts on stumps, from seedlings that survive the logging operation, or from seeds that germinate after the harvest. If natural regeneration is delayed longer than desired, the area is planted or seeded.

Clearcutting systems are often used to manage fast-growing species that require a lot of light. Resulting stands are even aged because all the trees in an area are cut-and regenerated-at the same time. Clearcutting has become controversial in recent years because it has the potential to damage watersheds and because it tends to eliminate species of wildlife dependent on old growth trees. If clearcuts are kept small and the cutting interval is long enough, however, biological diversity may not be impaired.

Shelterwood

In shelterwood systems, the forest canopy is removed over a period of years, usually in two cuttings. After the first harvest, natural regeneration begins in the understory. By the time the second harvest is made, enough young trees have grown to assure adequate regeneration. Shelterwood systems favour species that are intermediate in tolerance to shade. Such systems are difficult to use successfully and are the least used of the three silvicultural methods described.

Multiple-Use Forestry

Gifford Pinchot, the first Chief of the U.S. Forest Service, was also this country's first professional forester. Pinchot advocated the use of forest resources-all resources, not just timber-for human benefit. Pinchot was a strong and charismatic leader, and his ideas helped shape the course of forestry in the United States. Pinchot had a vocal opponent in John Muir, a young naturalist from California who believed that public lands should be preserved rather than used. Eventually Muir and Pinchot became rivals for public approval. Oddly enough, there was no

loser in this early conservation battle. Muir's preservation ethic became embodied in the philosophy of the National Parks, and Pinchot's concept of wise use became the guiding principle of the National Forests. National Forests are still managed under the concepts of multiple use and sustained yield. The dominant uses of National Forests are considered to be wood, water, wildlife, forage (for domestic cattle and wildlife), and recreation. Extraction of minerals and other valuable products is also considered a legitimate use of National Forests.

Because Pinchot's philosophy left room for the "highest and best use" of a given area, the U.S. National Forests now include a wilderness system of more than 32 million acres (13 million ha) in which timber harvest is not allowed. Today it is generally recognised that most, if not all, non-destructive uses of forest are valid. Some areas may be set aside as parks; others for wildlife habitat or as wilderness. Still others will be managed for timber harvest or multiple benefits. Today, conflicts arise primarily over where these different uses will be dominant. In the National Forests, such decisions are made through a land-use planning process in which the public has ample opportunities for input and involvement.

FORESTRY RESEARCH

At the turn of the century, very little was known about the world's native forests or how to manage them. In the United States, foresters were quick to recognise the value of information about forests and a branch of research was established in the Forest Service in 1915. Early research was done primarily in support of reforestation efforts, but, as forestry grew in size and complexity, so did the research. Today, the USDA Forest Service has six regional experiment stations located in important forest regions. Each experiment station has several field laboratories generally with specialized assignments for a geographic region or a specific subject area, and numerous sites for field research. In addition, the Forest Products Laboratory in Madison, WI, serves as a nationwide centre for research and development of new technology relating to wood, including tropical woods. Two laboratories are dedicated exclusively to tropical forest research: the International Institute of Tropical Forestry in Puerto Rico and the Institute of Pacific Islands Forestry in Hawaii.

Research is vital for modern forest management, which is information intensive. Today's foresters require vast quantities of data and a knowledge of ecology: they must understand not only the parts of ecosystems but how different parts of the environment interact. Scientific investigations are conducted in support of all kinds of forestry activities: silviculture, forest insect and disease control, wildlife habitat management, fire prevention and control, range and watershed management, forest products utilisation, forest survey, reforestation, ecology, and economics.

ECOSYSTEM OF TROPICAL RAINFOREST

A tropical rainforest is an ecosystem type that occurs roughly within the latitudes 28 degrees north or south of the equator (in the equatorial zone between the Tropic of Cancer and Tropic of Capricorn). This ecosystem experiences high average temperatures and a significant amount of rainfall. Rainforests can be found in Asia, Australia, Africa, South America, Central America, Mexico and on many of the Pacific, Caribbean, and Indian Ocean islands. Within the World Wildlife Fund's biome classification, tropical rainforests are thought to be a type of tropical wet forest (or tropical moist broadleaf forest) and may also be referred to as *lowland equatorial evergreen rainforest.*Tropical rainforests can be characterised in two words: warm and wet. Mean monthly temperatures exceed 18 °C (64 °F) during all months of the year. Average annual rainfall is no less than 168 cm (66 in) and can exceed 1,000 cm (390 in) although it typically lies between 175 cm (69 in) and 200 cm (79 in). This high level of precipitation often results in poor soils due to leaching of soluble nutrients.

Tropical rainforests are unique in the high levels of biodiversity they exhibit. Around 40 per cent to 75 per cent of all biotic species are indigenousto the rainforests. Rainforests are home to half of all the living animal and plant species on the planet. Two-thirds of all flowering plants can be found in rainforests.

A single hectare of rainforest may contain 42,000 different species of insect, up to 807 trees of 313 species and 1,500 species of higher plants. Tropical rainforests have been called the "jewels of the Earth" and the "world's largest pharmacy", because over one quarter of natural medicines have been discovered within them. It is likely that there may be many millions of species of plants, insects andmicroorganisms still undiscovered in tropical rainforests.

Tropical rainforests are among the most threatened ecosystems globally due to large-scale fragmentation due to human activity. Habitat fragmentation caused by geological processes such as volcanism and climate change occurred in the past, and have been identified as important drivers of speciation.

However, fast human driven habitat destruction is suspected to be one of the major causes of species extinction. Tropical rain forests have been subjected to heavy logging and agricultural clearance throughout the 20th century, and the area covered by rainforests around the world is rapidly shrinking.

HISTORY

Tropical rainforests have existed on Earth for hundreds of millions of years. Most tropical rainforests today are on fragments of the Mesozoic era supercontinent of Gondwana. The separation of the landmass resulted in a great loss of amphibian diversity while at the same time the drier climate spurred the diversification of reptiles. The division left tropical rainforests located in five major regions of the world: tropical America, Africa, Southeast Asia, Madagascar, and New Guinea, with smaller outliers in Australia. However, the specifics of the origin of rainforests remain uncertain due to an incomplete fossil record.

TYPES OF TROPICAL RAINFOREST

Several types of forest comprise the general tropical rainforest biome:

- Lowland equatorial evergreen rain forests are forests which receive high rainfall (more than 2000 mm, or 80 inches, annually) throughout the year. These forests occur in a belt around the equator, with the largest areas in the Amazon Basin of South America, the Congo Basin of Central Africa, Indonesia, and New Guinea.
- Moist deciduous and semi-evergreen seasonal forests, receive high overall rainfall with a warm summer wet season and a cooler winter dry season. Some trees in these forests drop some or all of their leaves during the winter dry season. These forests are found in parts of South America, in Central America and around the Caribbean, in coastal West Africa, parts of the Indian subcontinent, and across much of Indochina.
- Montane rain forests, some of which are known as cloud forests, are found in cooler-climate mountain areas. Depending on latitude, the lower limit of montane rainforests on large mountains is generally between 1500 and 2500 m while the upper limit is usually from 2400 to 3300 m.

- Flooded forests, seven types of flooded forest are recognised for Tambopata Reserve in Amazonian Peru:
- *Permanently waterlogged swamp forest*—Former oxbow lakes still flooded but covered in forest.
- *Seasonally waterlogged swamp forest*—Oxbow lakes in the process of filling in.
- *Lower floodplain forest*—Lowest floodplain locations with a recognisable forest.
- *Middle floodplain forest*—Tall forest, flooded occasionally.
- *Upper floodplain forest*—Tall forest, rarely flooded.
- *Old floodplain forest*—Subjected to flooding within the last two hundred years.
- *Previous floodplain*—Now *terra firme*, but historically ancient floodplain of Tambopata River.

FOREST STRUCTURE

Rainforests are divided into different strata, or layers, with vegetation organised into a vertical pattern from the top of the soil to the canopy Each layer is a unique biotic community containing different plants and animals adapted for life in that particular strata. Only the emergent layer is unique to tropical rainforests, while the others are also found in temperate rainforests.

Forest Floor

The forest floor, the bottom-most layer, receives only 2 per cent of the sunlight. Only plants adapted to low light can grow in this region. Away from riverbanks, swamps and clearings, where dense undergrowth is found, the forest floor is relatively clear of vegetation because of the low sunlight penetration.

This more open quality permits the easy movement of larger animals such as: ungulates like the okapi (*Okapia johnstoni*), tapir (*Tapirus* sp.), Sumatran rhinoceros (*Dicerorhinus sumatrensis*), and apes like the western lowland gorilla (*Gorilla gorilla*), as well as many species of reptiles, amphibians, and insects.

The understory also contains decaying plant and animal matter, which disappears quickly, because the warm, humid conditions promote rapid decay. Many forms of fungi growing here help decay the animal and plant waste.

Understory Layer

The understory layer lies between the canopy and the forest floor. The understory is home to a number of birds, small mammals, insects, reptiles, and predators. Examples include leopard(*Panthera pardus*), poison dart frogs

(*Dendrobates* sp.), ring-tailed coati (*Nasua nasua*), boa constrictor (*Boa constrictor*), and many species of Coleoptera. The vegetation at this layer generally consists of shade-tolerant shrubs, herbs, small trees, and large woody vines which climb into the trees to capture sunlight. Only about 5 per cent of sunlight breaches the canopy to arrive at the understory causing true understory plants to seldom grow to 3 m (10 feet). As an adaptation to these low light levels, understory plants have often evolved much larger leaves. Many seedlings that will grow to the canopy level are in the understory.

Canopy Layer

The canopy is the primary layer of the forest forming a roof over the two remaining layers. It contains the majority of the largest trees, typically 30–45 m in height. Tall, broad-leaved evergreen trees are the dominant plants. The densest areas of biodiversity are found in the forest canopy, as it often supports a rich flora of epiphytes, including orchids, bromeliads, mosses and lichens.

These epiphytic plants attach to trunks and branches and obtain water and minerals from rain and debris that collects on the supporting plants. The fauna is similar to that found in the emergent layer, but more diverse. It is suggested that the total arthropod species richness of the tropical canopy might be as high as 20 million.

Other species habituating this layer include many avian species such as the yellow-casqued wattled hornbill (*Ceratogymna elata*), collared sunbird (*Anthreptes collaris*), African gray parrot(*Psitacus erithacus*), keel-billed toucan (*Ramphastos sulfuratus*), scarlet macaw (*Ara macao*) as well as other animals like the spider monkey (*Ateles*sp.), African giant swallowtail (*Papilio antimachus*), three-toed sloth (*Bradypus tridactylus*), kinkajou (*Potos flavus*), and tamandua (*Tamandua tetradactyla*).

Emergent Layer

The emergent layer contains a small number of very large trees, called emergents, which grow above the general canopy, reaching heights of 45–55 m, although on occasion a few species will grow to 70–80 m tall. Some examples of emergents include: *Balizia elegans*, *Dipteryx panamensis*, *Hieronyma alchorneoides*, *Hymenolobium mesoamericanum*, *Lecythis ampla* and*Terminalia oblonga*.

These trees need to be able to withstand the hot temperatures and strong winds that occur above the canopy in some areas. Several unique faunal species inhabit this layer such as the crowned eagle (*Stephanoaetus coronatus*), the king colobus (*Colobus polykomos*), and the large flying fox (*Pteropus vampyrus*). However, stratification is not always clear. Rainforests are dynamic and many changes affect the structure of the forest. Emergent or canopy trees collapse, for example, causing gaps to form. Openings in the forest canopy are widely recognised as important for the establishment and growth of rainforest trees. It's estimated that perhaps 75 per cent of the tree species at La Selva Biological Station, Costa Rica are dependent on canopy opening for seed germination or for growth beyond sapling size, for example.

ECOLOGY AND CLIMATE

Most tropical rainforests are located around and near the equator, therefore having what is called an equatorial climate characterised by three major climatic parameters: temperature, rainfall, and dry season intensity Other parameters that affect tropical rainforests are carbon dioxide concentrations, solar radiation, and nitrogen availability. In general, climatic patterns consist of warm temperatures and high annual rainfall. However, the abundance of rainfall changes throughout the year creating distinct wet and dry seasons.

Rainforests are classified by the amount of rainfall received each year, which has allowed ecologists to define differences in these forests that look so similar in structure.

According to Holdridge's classification of tropical ecosystems, true tropical rainforests have an annual rainfall greater than 800 cm and annual temperature greater than 24 degrees Celsius.

However, most lowland tropical rainforests can be classified as tropical moist or wet forests, which differ in regards to rainfall. Tropical rainforest ecology- dynamics, composition, and function- are sensitive to changes in climate especially changes in rainfall.

The climate of these forests is controlled by a band of clouds called the Intertropical Convergence Zone located near the equator and created by the convergence of the trade winds from the northern and southern hemispheres.

The position of the band varies seasonally, moving north in the northern summer and south in the northern winter, and ultimately controlling the wet and dry seasons in the tropics.

These regions have experienced strong warming at a mean rate of 0.26 degrees Celsius per decade which coincides with a global rise in temperature resulting from the anthropogenic inputs of greenhouse gases into the atmosphere. Studies have also found that precipitation has declined and tropical Asia has experienced an increase in dry season intensity whereas Amazonia has no significant pattern change in precipitation or dry season. Additionally, El Niño-Southern Oscillation events drive the interannual climatic variability in temperature and precipitation and result in drought and increased intensity of the dry season.

As anthropogenic warming increases the intensity and frequency of ENSO will increase, leaving tropical rainforest regions susceptible to stress and increased mortality of trees.

SOIL TYPES

Soil types are highly variable in the tropics and are the result of a combination of several variables such as climate, vegetation, topographic position, parent material, and soil age Most tropical soils are characterised by significant leaching and poor nutrients; however there are some areas that contain fertile soils.

Soils throughout the tropical rainforests fall into two classifications which include the ultisols and oxisols. Ultisols are known as well weathered, acidic red clay soils, deficient in major nutrients such as calcium and potassium. Similarly, oxisols are acidic, old, typically reddish, highly weathered and leached, however are well drained compared to ultisols.

The clay content of ultisols is high, making it difficult for water to penetrate and flow through. The reddish colour of both soils is the result of heavy heat and moisture forming oxides of iron and aluminum, which are insoluble in water and not taken up readily by plants.

Soil chemical and physical characteristics are strongly related to above ground productivity and forest structure and dynamics.

The physical properties of soil control the tree turnover rates whereas chemical properties such as available nitrogen and phosphorus control forest growth rates.

The soils of the eastern and central Amazon as well as the Southeast Asian Rainforest are old and mineral poor whereas the soils of the western Amazon (Ecuador and Peru) and volcanic areas of Costa Rica are young and mineral rich.

Primary productivity or wood production is highest in western Amazon and lowest in eastern Amazon which contains heavily weathered soils classified as oxisols. Additionally, Amazonian soils greatly weathered, making them devoid of minerals like phosphorus, potassium, calcium, and magnesium, which come from rock sources.

However, not all tropical rainforests occur on nutrient poor soils, but on nutrient rich floodplains and volcanic soils located in the Andean foothills, and volcanic areas of Southeast Asia, Africa, and Central America.

Oxisols, infertile, deeply weathered and severely leached, have developed on the ancient Gondwanan shields. Rapid bacterial decay prevents the accumulation of humus. The concentration of iron and aluminum oxides by the laterisation pro cess gives the oxisols a bright red colour and sometimes produces minable deposits. On younger substrates, especially of volcanic origin, tropical soils may be quite fertile.

Nutrient Recycling

This high rate of decomposition is the result of phosphorus levels in the soils, precipitation, high temperatures and the extensive microorganism communities. In addition to the bacteria and other microorganisms, there are an abundance of other decomposers such as fungi and termites that aid in the process as well. Nutrient recycling is important because below ground resource availability controls the above ground biomass and community structure of tropical rainforests.

These soils are typically phosphorus limited, which inhibits net primary productivity or the uptake of carbon. The soil contains tiny microbial organisms such as bacteria, which break down leaf litter and other organic matter into inorganic forms of carbon usable by plants through a process called decomposition.

During the decomposition process the microbial community is respiring, taking up oxygen and releasing carbon dioxide. The decomposition rate can be evaluated by measuring the uptake of oxygen. High temperatures and precipitation increase decomposition rate, which allows plant litter to rapidly decay in tropical regions, releasing nutrients that are immediately taken up by plants through surface or ground waters.

The seasonal patterns in respiration are controlled by leaf litter fall and precipitation, the driving force moving the decomposable carbon from the litter to the soil. Respiration rates are highest early in the wet season because the recent dry season results in a large percentage of leaf litter and thus a higher percentage of organic matter being leached into the soil.

Buttress Roots

A common feature of many tropical rainforests is the distinct buttress roots of trees. Instead of penetrating to deeper soil layers, buttress roots create a wide spread root network at the surface for more efficient uptake of nutrients in a very nutrient poor and competitive environment.

Most of the nutrients within the soil of a tropical rainforest occur near the surface because of the rapid turnover time and decomposition of organisms and leaves. Because of this, the buttress roots occur at the surface so the trees can maximize uptake and actively compete with the rapid uptake of other trees.

These roots also aid in water uptake and storage, increase surface area for gas exchange, and collect leaf litter for added nutrition. Additionally, these roots reduce soil erosion and maximize nutrient acquisition during heavy rains by diverting nutrient rich water flowing down the trunk into several smaller flows while also acting as a barrier to ground flow. Also, the large surface areas these roots create provide support and stability to rainforests trees, which commonly grow to significant heights. This added stability allows these trees to withstand the impacts of severe storms, thus reducing the occurrence of fallen trees.

FOREST SUCCESSION

Succession is an ecological process that changes the biotic community structure over time towards a more stable, diverse community structure after an initial disturbance to the community. The initial disturbance is often a natural phenomenon or human caused event. Natural disturbances include hurricanes, volcanic eruptions, river movements or an event as small as a fallen tree that creates gaps in the forest. In tropical rainforests, these same natural disturbances have been well documented in the fossil record, and are credited with encouraging speciation and endemism.

4

Forest Structure

Supporting the processes which create and maintain the forest is our job as sustainable forest farmers. Soils are a fragile resource in the cold/dry climate of the southern Blue Mountains. Forest pests and their predators are impacted by human activity which increases pest numbers. Restoration forestry returns and mimics the natural processes which entomologist Torolf Torgerson refers to as "the forest's immune system".

Forest structure is constantly changing in accordance with its internal pattern of growth and mortality and in response to external forces. Even in natural forests, stands may be 'felled' by natural events, such as storms and fires, whilst stands change between events at a measured pace as individual trees grow, compete, dominate and die. The events that disturb the apparent stability and orderly growth of the forest occur frequently enough to form a permanent influence on forest structure and composition.

This chapter reviews these disturbances and describes their influence on forests at various spatial and temporal scales using mainly European and North American examples. The 'ordinary' growth of undisturbed forest is considered. The disturbance regime of any particular forest is to some extent a reflection of climate. Whilst most disturbances are localized, some act on a grand scale, influencing forest patterns over an entire landscape. Oliver and Larsen (1996) give a comprehensive account of forest growth and change. Peterken (1996) provides an illustrated summary for the temperate region that is related to conservation issues.

DISTURBANCE AND NUMEROUS OF FOREST

When a forest that has stood, apparently unchanging, for as long as people can remember is blown down overnight there is little doubt that the forest has been 'disturbed'. The change is abrupt, the event is discrete and the effect on the forest is a complete discontinuity with the previous state of affairs. Although there is always a degree of air movement, it is possible to define, quantify and delimit storms in space and time, and thus to count and study their frequency, pattern and impacts.

When the deer in a forest become so numerous that they browse down all seedlings and basal shoots, the state of the forest is clearly being changed but it is more difficult to see this as a disturbance. The changes are gradual, and in any case herbivores are a natural component of the forest. Abrupt increases or decreases in populations may have immediately discernible effects that can be dated, but more often populations fluctuate over years. In this case the impact of the deer is chronic, a continuously operating site factor, rather than a discrete event.

When a large canopy tree in an old-growth forest becomes rotten at the core and hollowed out by fungi, it is likely to collapse in a breeze that would have offered no threat when it was sound.

The chance that a breeze will disintegrate a large tree increases when the branches are covered in ice, the crown is in full leaf or the soil has been soaked by weeks of rain. If one looks at the forest as a whole there will always be a few trees collapsing in this way each year, part of the perpetual cycle of regeneration, growth, death and decay. From this perspective, such mortality can hardly be regarded as a disturbance without destroying the meaning of the word. However, from the perspective of the patch of ground under that tree, the sudden disintegration of a tree that has stood for perhaps 300 years is indeed an event; at this smaller scale it can readily be regarded as a disturbance, which presents an opportunity for regeneration.

The cause of a disturbance is less certain the closer one looks. In the previous example, the cause can be the great age of the tree, the presence of fungi, the physical instability created by the hollow trunk, the wet soil or the burden of ice, or the breeze that finally brought it crashing to the ground.

There is even ambiguity in the cause of a massive blowdown: whereas the wind is an obvious cause, the fact remains that the stand would not have blown down if it had been young rather than mature, *i.e.* age is also a 'cause'. Thus, each disturbance has a chain or network of causation, and the kind, degree and scale of its impacts are determined by a multiplicity of factors. Put another way, a moderate gust of 50 km h^{-1} will disturb one stand but leave another unaffected, depending on the state of the stand and its circumstances.

Despite these uncertainties, the storm that blows a whole forest down overnight is obviously a disturbance, or is it? From the point of view of an individual person living near that forest there is little doubt that this is the event of a lifetime, likely to cause as much disturbance to that person's peace of mind as it does to the forest's structure.

However, storms recur and the forest will again grow to maturity, but in 200-300 years time the forest may again be blown flat. Since 200-300 years is well within the natural span of many forest dominants, it can be argued that blowdowns are a continuing factor in the life of the forest and the evolution of the species it contains. Events that count as disturbances on one time-scale may be ongoing ecological factors on another.

These points are made not to confuse but to emphasize that ecological disturbances are rarely, if ever, discrete events with sharp boundaries and an obvious single cause. Rather, they are often ill-defined in space and time, and the proximal cause may be insignificant in comparison with the multitude of contributory factors. Pickett and White (1985) have provided a definition of disturbance that incidentally conveys some of the uncertainty: 'a relatively discrete event in time that disrupts ecosystem, community or population structure and changes resources, substrate availability, or the physical environment'. Attempts have been made to distinguish catastrophic disturbances from ordinary disturbances; indeed there is a clear difference between a stand-destroying storm and a wind that snaps a few branches, but the difference is just a matter of degree. Whether one is considering a single kind of disturbance or the cumulative load of disturbances sustained by a single forest, we are dealing with a continuum.

FOREST STRUCTURE AND PATTERNS

The interactions between disturbances and forest trees generate a complex range of forest structures and patterns. Perhaps the best route through this complexity starts with two contrasting stand structures (for descriptive purposes a stand is taken to be a compact patch of woodland of about 2 ha).

1. *Even-aged:* stands are even-aged if the canopy trees all started growth in approximately the same year. A limited age spread of perhaps 10 years is permissible within the definition. Such stands originate after a stand-destroying disturbance or an abrupt change in land use. Colonizing trees may take longer than 10 years to fill the available space, but as the stand develops its age range narrows: latecomers will have smaller, lower crowns than those that started growth immediately and this disadvantage ensures that their mortality rate is greater.
2. *Mixed-age:* stands comprising a wide range of age-classes, from old trees to saplings, intimately intermixed. Such stands tend to form a mosaic of even-aged groups at the scale of one or two canopy trees. They function by a process of relatively constant gap formation, in which gaps formed by the loss of canopy trees or major branches are filled by groups of saplings. These groups compete amongst themselves as they grow, so that eventually only one or two trees become established in the canopy, thereby achieving a 1:1 replacement of the original gap-forming tree. Such stands can only survive where there has been no catastrophic

disturbance within the adult lifetime of the oldest canopy trees.

Even-aged stands develop through a sequence of stages, which are explaine. In effect, they become steadily more mixed-age as they develop, initially through recruitment in the underwood and eventually through mortality of canopy trees and replacement with individuals initiated later. In the absence of catastrophic disturbance, an even-aged stand eventually becomes mixed-aged, a stage described as 'shifting-mosaic steady-state' by Bormann and Likens (1981) or 'old-growth' by Oliver and Larsen (1996)

This relatively simple model is complicated by non-catastrophic disturbances. Gaps created by storms, for example, are often irregularly patchy. Canopy disturbance created by droughts may take the form of an irregular thinning. Likewise, fires of intermediate intensity commonly leave a fairly even scatter of large trees. Regeneration follows both patterns of disturbance, so that moderate disturbances eventually generate both small-scale patches of younger growth within a hitherto even-aged stand and two-storied stands in which the survivors from the previous generation persist in a matrix of younger growth.

Partial disturbances give rise to stands intermediate between the mixed-aged and even-aged stands defined above. They are mixed in the sense that different age-classes are mixed together at a small scale, even at canopy level, but contain distinct generations, *i.e.* concentrations of individuals within narrow age bands. Jones (1945) recognized that strong even-aged elements within mixed-age stands were common within virgin temperate forests.

At the stand scale, a plot of the age (and often also the size) of trees demonstrates pulses of recruitment, and it may be possible to partition the stand into two to four more-or-less distinct cohorts. In truly mixed-age stands, the age and size distributions take a negative exponential form. In stands with strong even-aged elements, the relationship remains a negative exponential but with peaks and dips. Thus, for example, the size distribution of *Quercus mongolica* in virgin old-growth in Hokkaido showed a fairly smooth, continuous, J-shaped distribution, even though the population comprised five distinct generations spanning 500 years.

A second complication arises from the tendency of gaps to enlarge once they have formed. For example, windthrow gaps within coniferous forests become foci for populations of the beetle *Ips typographies,* which then colonize and kill trees standing by the gaps. In forests of all kinds, once a gap has been opened by wind, neighbouring trees are more exposed to subsequent winds and thus more likely to be blown down. As gaps enlarge, so they tend to coalesce, until the stand ceases to be a canopy punctuated by discrete gaps and becomes an irregular scatter of fully grown trees in a matrix of shorter, and usually younger, individuals. European forest ecologists, led by Liebundgut (1959), have not recognized gaps as distinct features but have attempted to represent the structure of old-growth stands as a patchwork of different stages of stand development. Each stage is defined as a distinct spatial arrangement of trees of different sizes and ages and is represented as a developmental link between two or more other stages, the whole representing a cycle of stand development with several pathways. Several different versions of the classification of stages and the dynamic cycle have been published. Maps have been prepared to show the patchwork of stages in particular forests.

In practice, neither gaps nor the Liebundgut cycle fully represents the complex structure and dynamics of undisturbed old-growth. Gaps are readily defined and delimited in the early stages of canopy break-up but become far less distinct as break-up progresses and the stand becomes multi-layered. Furthermore, most gaps are small, so that the question of what counts as a gap is crucial in any quantification, especially in stands where the crowns of canopy trees rarely touch, let alone overlap, for example *Pseudotsuga menziesii* old-growth.

In the case of the Liebundgut stages, the patches so clearly defined and mapped on paper are found on the ground to be ill-defined and poorly delimited. Intermediate conditions are common. Each patch of a particular stage is seen to be internally heterogeneous, *i.e.* to be a micropatchwork of different structures. The stages and maps based on them represent reality as perceived at a particular scale: both the classification of stages and the maps would have been quite different if a smaller scale had been adopted.

In so far as they can be delimited, gaps have internal structure. The distribution of dead wood is generally irregular, for example a gap created by the fall of a single tree has an accumulation of crown branches and debris on one margin, leaving the rest of the gap with a clean floor. In many forest types the underwood of shrubs and groups of advance regeneration survives gap formation: its original irregularities will be increased by some flattening as the canopy tree falls.

In larger gaps it is common for one or two subcanopy trees to survive, thereby retaining a light shelterwood within the gap. Irrespective of size, gaps have an internal geometry which ensures that in the Northern Hemisphere the northern side receives more direct sunlight than the southern side and that the eastern side becomes drier in afternoon sun than the western side, which receives direct sunlight during the more-humid mornings.

Furthermore, light intensity is greater at the centre of gaps than at the edges. In general, larger gaps permit a larger representation of shade-intolerant species. Internal heterogeneity created by fallen trunks and branchwood generates irregularities in browsing intensity and small-scale shading patterns, and thus also contributes to the variety of regenerants.

Both gaps and the representation of stages have been used to quantify canopy turnover. Gaps are formed, or canopy space is vacated, at about 1 per cent per annum. The rate varies from year to year and on a scale of decades but where it has been measured in different forests, gaps have formed at 0.5-2 per cent annually. This implies a residence time of about 100 years for trees that reach the canopy. Where the extent and duration of growth stages have been quantified, it has been possible to compute the duration of the growth cycle at some 300 years and to determine the degree to which the age distribution of patches departs from the theoretical steady state.

In practice, individual trees live far longer than 100 years. Firstly, trees take several decades to reach the canopy, particularly those shade-tolerant species that remain virtually static when they are heavily shaded. Even fast-growing light-demanding species take 30 years or so to reach canopy height. Secondly, some canopy species are inherently short-lived, leaving space for other species to reside longer in the canopy. Thirdly, gap formation is not absolutely correlated with age of trees, so that some individuals live far longer than the majority. In temperate forests, individuals of oak, tulip, hemlock and lime regularly achieve ages of 300-500 years or more.

When forest structure is considered at a landscape scale we see that the fundamental difference between even-aged and mixed-age stands is a matter of scale. A forested landscape made up entirely of mixed-age stands has the same size distribution and age-class distribution as a landscape covered in even-aged stands, each of a different age. It is only when the forest is considered at a stand scale that even-agedness is perceived. A landscape of even-aged stands is usually mixed-aged in aggregate, *i.e.* the various stands comprising the landscape represent a range of ages. Thus, at a very large scale all forested landscapes are mixed-aged, although the scale of their even-agedness varies from groups no larger than a single canopy tree up to patches covering several thousands of hectares. Some historic fires have burned over whole counties, leaving patches of hundreds of thousands of hectares that would have regenerated as a massive even-aged stand, for example the widespread 500-year age-class of *Pseudotsuga menziesii* in the Cascades.

Variation in the scale of the patchwork has several consequences for forest composition and structure. A very small-scale patchwork implies a fairly steady rate of gap creation when computed at a large scale, which in turn implies that forest structure will remain more or less constant.

A large-scale patchwork implies that at moderate scales (say individual catchments of first- to fourth-order streams) forest structure can change widely with time according to the irregularities of disturbances, for example Yellowstone. Frelich and Lorimer (1991) used growth stages and transition times between each stage to compute the state of the landscape in a region of moderate disturbance, and found that most of the forest remained old-growth for most of the time.

FOREST CANOPY STRUCTURE AND ITS ECOLOGICAL FUNCTIONS

Wind river slide Forest structure, the spatial arrangement of both live and dead tree and plant material, strongly influences the spatial patterns of forest co-habitants due to the fundamental organization of food, shelter, and space. Structure is closely linked with composition, age, and history, including disturbance. Components of forest structures are leaves, reproductive parts, branches, trunks, butts, roots, and dead wood, and the airspace among them. G. Parker wrote an excellent review and synthesis in the 1995 Forest Canopies text, which I use as the basis for this lecture. I summarize the general structural features and microclimates peculiar to forest canopies, compare how some forests differ in these characteristics, and explore how structure affects microclimate. I focus on structure and environment of closed, continuous forests.

Several themes: 1) forest canopy structure has been poorly defined, and rarely represented to allow cross-site comparison; 2) measurements of microclimate usually focus on mean values, at a few locations and short time scales - variation has not been assessed; 3) studies on the relation of structure and function are uncommon. Lots of theory, but little measured.

DEFINITIONS OF CANOPY STRUCTURE

Canopy definitions Canopy is combination of all leaves, twigs, and small branches in a stand of vegetation; the aggregate of all the crowns. The canopy is a region as well as a collection of objects. Forest canopy structure is the organization in space and time, including the position, extent, quantity, type and connectivity of the aboveground components of vegetation. It is often useful to consider the open spaces between canopy elements and the atmosphere contained within and between crown as part of the canopy.

Terms Many terms have been used in the past: Physiognomy for shapes of individual crowns; architecture for growth patterns and forms of stems; organization for statistical distribution of canopy components; texture to the sizes of crown units composing the overstory. Units of canopy structure: usually, crowns of trees, but ultimately, is leaves and twigs. But often deal with statistical distribution of millions of leaves. Many scales are evident: foliage may be climbed or clustered; crowns are grouped into stands; stands into landscapes. Descriptors Many descriptors: eg. Maximum tree height (hmax) or mean height (h), number or biomass density of elements (stems ha-1); canopy cover (fraction of sky not covered by canopy); or leaf area index, ratio of total one-sided leaf area to projected ground area (LAI, m2/ m3). Less common are specifying 3-D organization of canopy elements. Almost always described by mean conditions, not an assessment of variation.

Quantification of Canopy Structure

White profile diagram For many years our view of forests has been two-dimensional (2-D) and the structure and dynamics of forest ecosystems have been studied and analyzed using data expressed as stem maps that indicate the X and Y coordinates of individual trees along with species and size information (*e.g.*, diameter at breast height). "Stand structure" usually referred to the species composition and tree size distribution of forest stands, and sometimes included information on the 2-D spatial distribution of trees and tree height. Because humans are confined to the ground, we have been unable to study in detail the 3-D structure of forests including the forest canopy. Instead, we inferred tree growth from diameter increment at breast height, and competitive interactions among trees from analysis of the relationship between horizontal spatial distribution and diameter growth or mortality patterns. However, we have always known that the dynamics of forest ecosystems, including growth and competitive interactions among individual trees, occur in the canopy and not at breast height. Canopy access tools have provided the means to quantify canopy structure in a more 3-d way.. Early characterizations of canopy structure were caricatures of the crowns of larger trees of forests - profile diagrams. Oldeman and others did more sophisticated images. But they tend to reflect only the peculiarities of the plot and provide little quantification of the vertical organization of the whole.

Leaf distribution – jess Whole canopy Lz (leaf-height density) can be estimated by assembling structural measurements of individual crowns made from observations from the grown or harvested stems. Lots of measurements in crop plants, very few for whole forests. This shows vertical stratification of leaf distribution for an eastern deciduous forest, gathered with crane data. Fish eye Optical point measurements more common. Hemispheric fisheye photography uses potential light environment at a point to assess canopy structure. Several devices now estimate LAI from in-canopy light measurements, relying on Beer's law and the absorbance of light.

Remote sensing has done a lot. Use the spectral quality of canopy light, the ratio of red/far red light transmitted through the canopy. Several measures calculated from combining different reflectance bands relate to the amount of green canopy biomass, such as the "normalized difference vegetation index (NDVI) or "greenness" index. Index takes advantage of the strong reflectance in the near infrared but weak reflectance in the red wavelengths of green canopies. Other remote sensing techniques use a similar approach, with sensors of different wavebands on several satellite platforms. (LIDAR) Also from a profiling airborne laser - to sense the elevation of both the ground and canopy, yielding the contour of canopy height over a long transect. So advances in quantifying and visualizing canopy structure is growing in sophistication.

Old growth messy The term "structural complexity" was coined as an integrative concept to represent the complex 3-D structure of old-growth forests. However, many of the measures and indices used to characterize structural complexity such as species composition, tree-size distribution and abundance of snags and woody debris are derived from 2-D, ground-based measurements. Ishii young old forest One of the objectives for developing indices of structural complexity is to quantitatively distinguish old-growth forests from younger stands and to establish criteria for enhancing old-growth forest structure in managed stands for conservation purposes, such as creating wildlife habitat. In order to meet such management objectives, we must integrate various aspects of forest structure including canopy and below-ground processes into future ecosystem management strategies.

VERTICAL ORGANIZATION

Canopy elements can be non-uniformly distributed with height. This pattern derives from species differences in growth form and shade tolerance, and stand developmental stage. Jess species distribution At its simplest, we can use an example from mixed deciduous forest. Vertical sorting of species leaf area. This is a foliage height profile for an Appalachian mixed species forest, giving the percentage of leaf area for the major species.

More generally, the vertical organization of microclimate, structures, and biota in forests is a recurring theme in scientific investigations and forest description. For example, Leaf characteristics such as size, shape, mass, inclination, chlorophyll content, N content, and photosynthetic capacity vary within a tree crown along gradients of light and other microenvironmental conditions so that the individual plant maximizes carbon gain. Vert strat definition One aspect of this is the concept of vertical stratification. Stratification proposes predictable vertical separation of canopy components such as forest leaves and other structures, species, or individual organisms into distinct horizons, layers or gradients.

Long history of vertical organization. Early naturalists, such as Allee (1949) saw stratification in the big picture, as a general phenomenon of animal and plant communities in aquatic (freshwater lakes, oceans, inter-tidal zones) and terrestrial (grasslands, deserts, forests) environments. Smith (1973) concluded that stratification optimizes light utilization, CO_2 concentrations, pollination and dispersal, reduces predation on flowers, fruits and leaves, and increases structural integrity of the forest. Strict adherence to the proposition of stratification has been criticized as too limiting when in fact there are ecological gradients in three dimensions from forest floor through the canopy that are complex mosaics of biota, microclimate, gaps, and the growth, mortality, and development of trees.

Horizontal Variation

Light gap Forests are not spatially uniform, but are horizontally heterogeneous at various scales. Most work has dealt with foliage-free spaces. Light gaps are holes in the canopy extending to the forest floor that permit the penetration of unscattered light (Runkle and Canham). Gaps of various sizes originate from a variety of causes. Lots of work done on this, as it influences bird flight, regeneration, seed germination. Crown shyness In some forests, individual subcrowns and crowns are clearly separate, with vegetation free borders in between.

This "crown shyness" is most common in single-species and single-cohort stands, and on windy sites. Probably maintained by wind-induced abrasion between adjacent crowns. "Crown asymmetry", where the crown centre is offset from the stem base, results from plasticity in the directional growth of tree crowns. Broad-leaved trees show great plasticity in crown growth that allows trees to grow towards open areas of the canopy and avoid competition from neighboring crowns. Crown asymmetry can buffer negative effects of crown competition, reducing tree mortality and increasing the mean and variation in tree size within stands. Coniferous trees have less plasticity in crown form and directional growth of the crown.

Temporal Changes

Light env over year Canopy structure changes seasonally in all forest, but is most dramatic in deciduous stands. But even evergreen forests, the quantity of leaf area varies over the year. Gen stand development More substantial changes occur on a succession time scale. Total amount stabilizes, but its vertical and horizontal distribution change slowly. Bob vp stands This difference has been shown through comparative studies of 2-D stand structure. At the individual tree level, can be seen. Jess images of stand ages But at the stand level, can also be quantified. Species composition and tree size distributions become more diverse with increasing stand age and specific structural elements such as large, old trees and snags characterize older stands

Deterministic Processes that Drive Development of Canopy Structure

In early stages of stand development, deterministic processes such as timing of establishment following disturbance, height-growth rate, and crown interactions determine canopy structure. In mixed-species natural forests, differences among species in timing of establishment and initial height-growth rates result in vertical stratification of species within the canopy. Early-successional, fast-growing species can establish soon after disturbance, reach the upper canopy, and dominate during early stages of succession.

The vertical development of canopy structure with increasing stand age can be inferred by comparing stands of different ages, *i.e.*, a chronosequence approach. Bob vp leaves In Douglas-fir and western hemlock forests of the Pacific Northwest Coast of North America, development of vertical canopy structure during early-successional stages begins with the dominance of the fast-growing, pioneer species, Douglas-fir, in the upper canopy of mixed-species stands. In late-successional stands, canopy height reaches over 70 m. Douglas-fir continues to dominate in the upper canopy as more shade-tolerant species such as western hemlock and western red cedar invade the mid- to lower-canopy.

This results in the development of a deep, continuous canopy comprising both early- and late-successional species As forests mature, canopy height reaches maximum and species differences in height-growth rate become less important in determining structural development. Species with similar levels of shade-tolerance will occupy similar positions in the canopy. Late-successional species that can reach the upper canopy eventually catch up and take the place of less shade-tolerant, early-successional species in the upper canopy. However, in regions where there are few physical limitations to canopy height (*e.g.*, typhoons and poor soil conditions), physiological limitations to height growth and crown expansion may reduce competition among the tallest trees, allowing early successional species to coexist in the upper canopy with the late-successional species.

For example, in the old-growth Douglas-fir–western hemlock at the Wind River Canopy Crane Research, mild climate and abundant rainfall allow canopy trees to reach heights over 60 m. In this forest, upper-canopy trees (tree height > 40 m) of all species have attained maximum crown size and show very little crown expansion growth.

In the absence of crown competition, long-lived pioneer species such as Douglas-fir may be able to coexist with the late-successional species in the upper canopy.

STOCHASTIC PROCESSES AND DEVELOPMENT OF STRUCTURAL COMPLEXITY

With increasing stand age, stochastic or random processes such as mortality of individual trees that create gaps in the canopy and small-scale disturbances that cause damage and die-back of crowns play increasingly important roles in creating structural complexity of the forest canopy. As trees reach maximum size, limitations to crown expansion in the upper canopy enhance structural complexity of the canopy surface. In Douglas-fir–western hemlock forests, gaps created in the upper canopy after individual tree mortality are not filled by neighboring trees and the upper-canopy surface becomes increasingly heterogeneous with increasing stand age. Many trees in old-growth forests show evidence of past damage and regrowth, such as forks and crooks in the main stem. In tropical forests and monsoon regions that are frequently affected by windstorms and typhoons, crown damage is commonly observed. oughts, fungal infections, insect outbreaks, and forest fires can also cause defoliation and die-back of the crown and are often followed by regrowth. Stochasticity of small-scale disturbances followed by regrowth of the crown adds variability to the otherwise deterministic architecture of trees and enhances structural complexity of the forest canopy.

Reiteration contributes to prolonging tree longevity by reproducing dead and dying crown components. Many long-lived, late-successional species can reiterate architectural units and maintain the crown, in contrast to short-lived early-successional species which tend to have less ability to reiterate crown components. Reiteration of various architectural units ranging from shoots and twigs to entire branches and vertical axes (reiterated trunks) has been observed in some large, old trees of long-lived species, including redwoods and Douglas-fir. Understory trees of European beech, a shade-tolerant, late-successional species, maintain the crown by means of reiteration when growth is suppressed due to limited light conditions. In old Douglas-fir trees, reiteration enhances structural complexity of the crown by increasing branch size variability. Reiteration also contributes to enhancing structural complexity by creating specific structural features such as reiterated trunks, fan-shaped clusters of epicormic branches, and platform-shaped forks within branches, making the crown of each tree "highly individualistic and irregular

Marbled murreletWhy is this important? These developmental processes of canopy structure drive and enhance various ecological functions such as community dynamics and stand productivity. Spotted owl Development of structural complexity enhances biodiversity of forest ecosystems by creating specific structural elements that provide food and habitat for other organisms.

Forest Micro-Environmental Gradients

Mv cloud Canopy microclimate is ultimately determined by the stand macroclimate; the rhytms of change above and within the forest are set by the cycles of annual and diurnal heating and by the movements of air masses and clouds. Very short term events (time scale of less than a minute) are important in exchange process and ventilation of canopy layers. Some environmental variables are influenced by broadscale canopy features (wind), whereas others (light) are depended on the local arrangements of elements. The vertical pattern of microenvironmental conditions within forests is important because it influences the distribution of forest biota, behaviour of vertebrates, the development and growth of tree structures, amount of gas exchange and water release by forest leaves, infection potential for numerous tree parasites, invasiveness of lianas into tree crowns, growth and productivity of epiphytes, biological activity levels of microbes, and numerous other aspects of ecosystem function.

The forest influences microclimate, and microclimate influences how and where the forest will grow, so that the interaction of the two results in measured patterns in a dynamic state.

The three-dimensional microenvironment (light, humidity, temperature, and wind) of forests is spatially heterogeneous based on composition and structure of the forest. Wind distribution Wind, for example, has higher peaks in the upper canopy, than down near the forest floor. In general, wind is rapidly decelerated in the layer just above the forest The velocity profile in this layer is commonly described as though the wind were reacting to a rough surface displaced above the ground by a distance d. This is equivalent to the mean height of the momentum absorption, with a gradient controlled by the roughness of the surface. Both the roughness and the displacement height of canopies depend on the amount and distribution of canopy material, and also on wind speed itself. Estimates are often based on the canopy height. In some stands, wind speeds do not decline monotonically with depth, but have a secondary max in the lowermost canopy levels. This is seen in stands of simple structure that lack vegetation layers at the bottom, or at forest boundaries, where winds may blow through for distances equal to several canopy heights.

The canopy acts as a filter of high frequency gusts, arresting small-scale fluctuations, but permitting the penetration of large eddies. The depth to which eddys penetrate depends on canopy density, strength of the eddy, stability of the canopy air column. Much of the total transport occurs during a small fraction of the time. Periods of relative quiescence are punctuated with gusts that can penetrate deeply. With high-frequency sensors, rapid vertical motions and temperature deviations may be observed almost simultaneously at several canopy levels.

Trace gases and particle concentrations have relatively weak gradients within the canopy, but mean concentrations are typically lower in the understory than in the overstory. For CO2, however, the active layer of the canopy is an enormous sink, and CO2 concentrations are often slightly depressed in the overstory during the daytime. The ground is a source due to root and soil respiration and decomposition. A pronounced co2 max develops in the understory late at night, especially under stable conditions.

Light is perhaps the most influential and complex of all canopy microclimatic variables. Most microenvironmental factors correlate with light intensity. Therefore, the effect of microclimate on plant physiology and growth as well as on animal behaviour and occurrence is difficult to separate from that of light alone.

This makes light a suitable universal indicator for just about every aspect of forest canopy ecology, but it is very difficult to measure in three-dimensional space. Parker proposed a vertical subdivision of canopies into three zones based on the patterns of the mean and variance of vertical light transmittance; bright, transition, and dim. The bright zone (upper canopy) is characterized by high transmittance and low variability, the transition zone (mid canopy) is where transmittance is most variable and the mean changes rapidly with height, while the dim zone (lower canopy) is characterized by low transmittance and variability. Based on work by Parker an age class sequence of Douglas-fir forests near the WRCCRF, the bright zone is wide and dim zone narrow in older stands due to the vertical differentiation of the canopy with age. Younger forests have a narrow bright zone and wide dim zone.

Canopy Soil Characteristics

Canopy soils are a particularly interesting habitat due to their patchy spatial distribution and vertical organization. Soils develop where there is a perch that allows the accumulation of litter, such as under and around epiphytes, on large branches, and in branch crotches. DOM is especially interesting because it can capture incoming nutrients bystoring them on the negatively charged sites, lots of recalcitrant nutrients,and support of invertebrates, and its ability to store water for a long time.

Several studies have examined differences between canopy and forest floor soil - In Monteverde, compared nutrients and pH. Also studied differences in soil temperature and soil moisture soil temp. Not much difference for temperature, but striking idfferences for moisture. Severe drydowns occur in the forest canopy, not the soil. This could have a role in determining the composition and abundance of vertical gradient. Soil moisture

FOREST SOILS AND ACIDIC INPUTS

Glaciers have advanced and retreated across Massachusetts many times over the past several million years. The last glacier retreated over 10,000 years ago. Our soils are the product of the interaction of vegetation and climate on what the glaciers left behind. Most upland, forested soils have developed out of what is called `glacial till,` that is, soils that were created and mixed by glaciers moving across the landscape. They are typically stoney, fine-textured soils on top of bedrock.

They may or may not have hardpan layers. We also have pockets of `glacial deposit` soils which were left by streams and rivers created by melting glaciers. These deposit soils are made of mixtures of sand and gravel and are found where there were edges of tongues of the retreating glaciers--mostly in river and stream valleys.

Many hillsides have scattered pockets and bands of these soils among the predominantly glacial till soils. There are also lacustrine and alluvial soils. These are both silty soils, the former from deposits on old lake bottoms and the latter from flood deposits along rivers and streams. Most of these soils are in agricultural use.

Sulphur emissions have increased since the start of the Industrial Revolution to the extent that 75 per cent of global emissions of S (total about $105 Mt year^{-1}$) come from anthropogenic sources. While emissions from natural sources are about evenly distributed between Northern and Southern hemispheres, 90 per cent of anthropogenic sulphur emissions come from sources in the Northern Hemisphere (about $70 Mt year-^{1}$ in the Northern Hemisphere and $7.7 Mt year^{-1}$ in the Southern Hemisphere.

Forest Decline

Forest decline in the Northern Hemisphere was increasingly observed during the 1970s and 1980s and was initially linked with the increasing acidity of rain, the latter caused by nitric and sulphuric acid formed from emissions from industry, cars, etc.

However, there are other sources of acidity such as dry deposition (including both direct gaseous absorption and particulate deposition), and the rates of dry deposition of SO_2 and NH_4^+ may exceed rates of deposition in rainfall. There are also many causes of forest decline other than acid deposition.

There is little doubt that acid rain (used here in a generic sense) is a cause of some severe forest declines, particularly when close to emission sources. The most extensive studies have been in Germany and the results can be simply summarized.

With increasing acidity, cation exchange capacity is reduced (leading to the loss of Ca^{2+} and Mg^{2+}) and the concentration of Al ions increases. Ca: Al and Mg: Al ratios in needles of *Abies alba* from healthy stands are greater than those in needles from stands that show symptoms of forest decline.

Lindberg *et al.* (1986) estimated that atmospheric deposition supplied 40 per cent of the annual requirement for wood production of S and 100 per cent of that for N. Continuing input of acid rain therefore produces imbalances in the ratio of N to cations in tree crowns that result in chlorosis and, if serious enough, death. However, for many forests cause and effect are not easy to link. Most forest soils are acidic, so there is the overriding problem of assessing increasing acidification of an already acidic medium; added to this is the difficulty of determining and assessing the sources of acid input.

Two themes show the complexity of the acid rain problem. Markewitz *et al.* (1998) analysed soil samples from the Calhoun Experimental Forest that had been archived over three decades.

They concluded that natural processes of acidification (increase in biomass and uptake of $[cations]_c > [anions]_c$ by plants, loss of $[cations]_c > [anions]_c$ by leaching, increase in soil organic matter, mineralization of N to NO_3-, microbial and root respiration, production of organic acids) accounted for 68 per cent of acidification over the 30 years, with acid deposition accounting for the remaining 32 per cent. With a threefold increase in SO_2 emissions for the eastern USA from 1960 to 1980, the capacity of the surface soil to adsorb $SO_4{}^{2}-$ was saturated but there was strong adsorption below 60 cm.

A result of major interest is that, from 1982 to 1990, the concentration of extractable $SO_4{}^{2}-$ in soils at 35-60 cm decreased significantly. This decrease was equivalent to a loss of charge of $2.2 kmol_c ha-^{1} year-^{1}$, and Markewitz *et al.* (1998) suggest that the decrease in concentration of extractable $SO_4{}^{2}-$ shows that the soils at Calhoun are responding (or recovering) to the decrease since 1980 in rates of both SO_2 emission and $SO_4{}^{2}-$ deposition.

For the intensively studied Hubbard Brook Experimental Forest, north-eastern USA, inputs of acid rain have caused large losses of cations (particularly Ca^{2+} and Mg^{2+}, together with K^+ and Na^+) in drainage waters. Since 1963, atmospheric emissions have decreased, resulting in decreases in concentrations of $SO_4{}^{2}-$ and the cations Ca^{2+}, Mg^{2+}, Na^+ and K^+ in bulk precipitation.

The loss of cations in drainage and the net uptake of cations by the forest together represent an 'ecosystem requirement' that must be balanced by inputs in bulk precipitation, from weathering and from release of cations from exchange sites in the soil. For Ca^{2+}, precipitation supplied 29 per cent and net release from soil 63 per cent of this required balance during the period 1940-55.

From 1976 to 1993, precipitation supplied only 12 per cent of demand and net release from soil increased to 79 per cent. Likens *et al.* (1996) therefore concluded that even with major reductions in emissions, the response of soil and streamwater chemistry (particularly the acid-base balance) will be significantly delayed relative to that expected from studies of inputs of acids (the bio-geochemistry of sulphur) alone.

Nitrogen Saturation

The second theme is called N saturation, the hypothesis that forests may become 'N saturated' as a result of too much N from acid deposition. Again, this is a concern for the Northern Hemisphere forests, for example in Europe , in North America and in Japan. Binkley and Högberg (1997) examined the hypothesis of N saturation in considerable analytical detail for forests in Sweden.

While S deposition rates in Sweden have decreased by as much as 50 per cent since 1980, the annual rates are still as high as 25 kg ha^{-1} along the south-west coast, decreasing to <12kgha^{-1} in the north. These authors also estimated the annual rates of input of N to the south-west coast as 8-12 kg ha^{-1} for NO_3–-N and 5-15 kg ha^{-1} for NH_4^+-N, decreasing in the north of Sweden to <2kgha^{-1} for both NO_3–-N and NH_4^+-N.

Despite these sometimes high inputs of N, the N is apparently accumulated since in only a few stands near the south-west coast do losses of N by leaching approach inputs of N in deposition. Thus the evidence is that Swedish forests are not N saturated, and forest decline is not linked with any condition of N saturation. The evidence from the application of N fertilizers is uncertain to say the least; some stands respond positively, some negatively and some not at all. Nevertheless, Binkley and Högberg (1997) concluded that 'the vast majority of conifer forests in Sweden still suffer from chronic N deficiency'.

Because of concerns about the acidification of forest soils a number of countries have initiated liming experiments, such as the Höglwald experiment in Germany, where 4tha^{-1} of dolomitic limestone was applied.

In reviewing these experiments, Binkley and Högberg (1997) concluded that 'the overall effect of liming of forest soils in the Nordic countries has been a general decrease in forest growth (of perhaps 10 per cent)'. A piece of jargon that has emerged from soil acidification and the N-saturation hypothesis is 'vitality fertilization' (or, even worse, 'vitalization fertilization'), a term that really means no more than the application of a fertilizer to ameliorate a deficiency (or, more ornately, to correct a nutrient imbalance). We agree strongly with Binkley and Högberg (1997): 'vitality fertilization' should not be used because it implies lack of vitality where, as in most forests, there is no symptom of decline.

Acid deposition is a problem for the Northern Hemisphere, simply as a consequence of the very much greater density of land area and population in the Northern than in the Southern Hemisphere.

While forest decline is caused by acid rain in many forests, in many other forests in both hemispheres decline cannot be linked with high inputs of acid. Wild (1993) summarized a number of causes of decline other than acid deposition.

- Natural acidification processes lead to accumulation of Al and a deficiency of Mg in old trees. This problem is exacerbated by an increase in N supply that causes the development of large crowns, placing demands on Mg. Where Mg is in short supply, Mg is translated from older foliage to developing foliage.
- The increased input of N produces softer or more palatable tissues, resulting in increased insect attack.
- Ozone is directly absorbed by leaves and is extremely damaging to structure and function; decline may therefore be caused by increased inputs of ozone.
- The 1970s and 1980s in Europe were characterized by hot dry summers and cold winters; these conditions might place stress on some species in some areas.
- Poor forest management.

In summary, the 1980s saw the widespread prediction of a general decline of European forests (the concept of *Waldsterben).* By the 1990s this view was greatly modified, particularly by evidence that the growing stock of European forests increased between 1971 and 1990 by 25 per cent and that growth of the forests increased by 30 per cent. Skelly and Innes (1994) concluded that 'the concept of a general forest decline is untenable', and that the evidence for air pollution causing forest decline 'away from known sources of pollution is extremely limited'.

LITTER AND SOIL ORGANIC MATTER

Litter and Litter Decomposition

There have been several syntheses of rates of accession and decomposition of litter in forests over the last two decades. Large amounts of litter fall from the above-ground parts of forests to the soil annually; this litter is a major determinant of nutrient cycles in forests. On a global scale, annual above-ground litter production in forests, woodland and wooded grassland is estimated at about 39 Pg of dry matter (1Pg= 10^{15}g), which is a little less than 50 per cent of total global litter production. Annual above-ground litter accession is highest in tropical, broadleaved, evergreen forests at 15300kgha^{-1} and as low as 130 kg ha^{-1} in boreal, needle-leaved, evergreen forests.

On a global scale, average litter turnover time is about 5 years for litter and about 13 years for coarse woody debris (usually classified as materials >7cm diameter). As decomposition of forest litter is an important component of global C cycling, the potential impact of climate change on C release from litter has attracted considerable attention over the last decade. In a recent review on a global scale of litter decomposition (measured as decomposition in the first year), Aerts (1997) concluded that climate (expressed as actual evapotranspiration) is a better predictor of the rate of litter decomposition than litter chemistry, although climate indirectly influences litter chemistry. However, litter chemistry became the best predictor of decomposition rates within discrete climatic regions, especially in the tropics and in the Mediterranean region where the ratio of concentration of lignin to concentration of total N was the best indicator of litter decomposition. The strong correlation between decomposition rates and the lignin: N ratio of litter has been corroborated in a wide range of Canadian forests where 73 per cent of the variance in decomposition was accounted for by a multiple regression of mean annual temperature, mean annual precipitation and the lignin: N ratio of the litter.

On a regional or global scale, the rate of litter decomposition in forests is predominantly controlled by climate (temperature and moisture, as shown in general terms by a strong negative correlation with latitude) and by the quality of litter in terms of how beneficial it is to the micro-bial community as a source of energy or nutrients.

High concentrations of nutrients relative to stored energy promote more rapid decomposition rates, so that C: N and lignin: N concentration ratios are commonly used as variables for litter quality. Across the generally mesic forested biomes, climate and litter quality seem to be the best indicators of the rate of litter decomposition. However, the value of litter quality as a determinant of the rate of litter decomposition does not hold across all biomes, and is poorly correlated with the rate of decomposition in desert regions.

While accession of exogenous N to forested land has raised concern for increased decomposition of litter, reports of effects on decomposition rates vary. For example, Kuperman (1999) reported increased rates of litter decomposition (and of N mineralization) in white oak stands of the lower midwestern USA following increased N deposition. However, in a study encompassing a wide range of sites, Prescott (1995) found that increased N availability following fertilizer addition or deposition did not alter rates of litter decomposition in forests.

It seems clear that while exogenous N may increase decomposition in some forests, especially over the first year following litterfall, it may also increase the lignin content of litter and reduce long-term decomposition rates in other forests. The reduced decomposition of N-rich litter may be due to the chemical formation of stable nitrogenous compounds from lignin by-products.

These N compounds are thought to form by chemical condensation reactions and may be highly resistant to biological degradation. Current evidence therefore suggests that an increased rate of N deposition decreases the rate of C release in the latter stages of decomposition and stimulates C storage, both in increased woody biomass and through soil humus formation and, thereby, sequestration of C as soil organic matter.

Below-ground inputs of C and nutrients in forests may become the dominant pathway of nutrient return, especially in cold-temperate climates. Berg *et al.* (1998) studied root litter decomposition in coniferous forests from northern Scandinavia to north-eastern Germany and found little correlation between the loss of mass in the first year and climatic variables when different forest types were considered together. When the *Pinus* and *Picea* stands were treated separately, stronger correlations emerged with initial concentration of P and July temperatures explaining 71 per cent of the variation in the rate of decomposition of root litter of *Picea* stands along the transect.

Soil-warming experiments in a northern hardwood forest, where the forest floor was heated with a cable, showed that the CO_2 flux from leaf litter increased exponentially with increasing soil temperature. These increased rates of CO_2 flux from litter decomposition may represent a short-term depletion of labile C with increasing temperature, so that the response diminishes over time.

However, if global warming causes increases in the rate of biomass production and in the rate of litterfall, the rate of litter decomposition will increase and increases in the rate of CO_2 emissions from forest litter will be sustained.

Forest Soils and Ccarbon Storage

The global cycling of C involves fluxes between the fossil C reservoir, the atmosphere, the oceans and the terrestrial biosphere. The role of the terrestrial biosphere in buffering or contributing to increases in CO_2 of the global C cycle is uncertain, partly because of the difficulty in predicting the likely impact of climate change, associated with rising atmospheric CO_2, on C flux.

Because forests and wooded lands cover about one-third (4100 million ha) of the land area of the earth, they are a significant component of that C stored in the terrestrial biosphere which is exchangeable with the atmospheric pool of C. The estimated 787 Pg of C in forest soils and associated peats is more than two-thirds of the total of 1146Pg of C held in forests.

About 33 per cent of forest soils are covered by coniferous forests in the high latitudes of Russia, Canada and Alaska. About 25 per cent of forest soils are covered by forests in the mid-latitudes of continental USA, Europe and Australia, while the low-latitude (tropical) forests of Asia, Africa and the Americas account for 42 per cent of forested land. More than half of the low-latitude forests are in tropical America. The amount of C stored in forest soils increases with northern latitude, so that low-latitude (tropical) forests contain about 27 per cent of the pool of soil C, increasing to about 60 per cent in soils of high-latitude northern boreal forests.

Thus the soils of high-latitude northern forests are an important C pool and any change in the climate or the management of these forests that diminishes the soil C pool will significantly affect global C storage. This applies equally to the tropical forests of lower latitudes, where soil C stocks are less than half those of the boreal forests, but where small increases in temperature will have a large proportional effect on metabolic activity.

5

Genetics and Improvement of Exotic Trees

The genetic improvement of exotic trees is based on the same general principles that are valid for all forest trees and has the same object, namely, increased production, both qualitative and quantitative, and the development of all those characters, such as plasticity, tolerance and resistance, which enable the maximum benefit to be obtained from planting trees.

However, the practical achievement of this improvement comes up against difficulties linked with the exotic origin of the trees. For example, knowledge of the biology, ecology and genetic variability of the species in its natural home is often very incomplete and hard to obtain in underdeveloped regions. The supply of reproductive material as seed, grafts or cuttings of known origin is also difficult to organize.

Rcscarch on the improvement of exotic trees should therefore be done both in the country of origin (where the research should include the study of the genetic variability, provenances, selection of seed stands and plus trees), and in the country of planting (acclimatization tests, experiments with provenances, hybridization and the formation of seed orchards). The current trends in wood consumption suggest the primary concentration of effort on those trees whose present and future economic value is well known.

IMPORTANCE OF EXOTIC TREES

The tremendous increase in the demand for wood, due to economic and industrial development, has brought about a revolution in forestry which traditionally has been linked with century-long rotations and a largely extensive type of production. Foresters have been forced to adapt themselves to the new industrial rhythm and to concentrate their efforts on transferring this rhythm to forest production. For this they must adopt new techniques and especially new means of production, in other words, new tree species. In certain regions the selection and improvement of the indigenous forest tree species has provided the means of production, but very frequently the local species did not have the required characters and it became necessary to introduce trees from other regions. This development has given us two new terms, namely "introduced" or "exotic" species. It is not necessary to provide examples of the introduction of trees; during the last few years the exchange of trees between one country and another has become almost as frequent and widespread as the exchange of agricultural plants.

It must first be emphasized that the adjectives "exotic" or "introduced" apply to a species grown outside its natural habitat. This definition leads one to consider as exotics, for phyto-geographic reasons, some species that are "politically" indigenous. This applies to European larch *(Larix decidua)* when taken from the Alps to the German plains; to the Corsican pine *(Pinus nigra* var. *calabrica)* planted in Metropolitan France; and to the western white pine *(Pinus monticola)* when it is transferred to the eastern United States.

Naturalists were the first to introduce new tree species, mainly from the eighteenth century onward, largely to establish botanical collections or to enrich parks and gardens. The nineteenth century, however, saw the commencement of the introduction of trees to fulfill the needs of forestry. To-day four basic reasons for introducing species are recognizable:

1. To enrich the local flora;
2. To obtain resistance to disease and other unfavorable environmental factors;
3. To exploit superior growth rate;
4. To obtain wood quality.

ENRICHMENT OF THE LOCAL FLORA

The present composition of the forest flora of a region is the result of natural evolution in relation to climatic variations throughout geological time (and especially the glaciations) and of the direct and indirect action of man.

Consequently, the forest flora often has a very different composition in regions which have similar ecological conditions. This applies especially to conifers which are the basis of the modern forest economy the distribution of which is extremely irregular. In tropical and subtropical zones, for instance, broadleaved trees are largely dominant, save for a few countries such as Mexico where there are also conifers. In other cases the local forest flora consists only of shrubby trees and bushes, even where the ecological conditions would permit the growth of large-size trees.

Thus the introduction of exotic trees can supplement the local flora and provide foresters with a new means of production.

RESISTANCE TO DISEASE AND OTHER UNFAVORABLE ENVIRONMENTAL FACTORS

The need for resistance to disease and other unfavorable site factors is often the chief incentive for introducing new species. The resistance of the

Japanese larch *(Larix leptolepis)* to the agent of larch canker *Dasyscypha (Trichocyphella)* willkommii which affects the European larch *(Larix decidua)*; the use of the Japanese and Chinese species of *Castanea* which are resistant to the ink disease of sweet chestnut and to Endothia parasitica; the plantations of *Picea sitchensis* which resists strong winds in the northern European lowlands; the use of certain species of *Acacia* and *Tamarix* in saline soils; all of these prove the importance of widespread testing and use of exotic species.

SUPERIOR GROWTH RATE AND PRODUCTION

The need for greater rates of growth normally provides the reason for preferring exotic to indigenous species. Two factors must be considered. First, the capacity of the species to grow to a large size; and secondly, rapidity of growth which makes it possible to introduce the rhythm of industrial output to wood production. Some species can produce a given volume of wood in half the time required by others. Moreover, if volume increment is high from the early years, the thinning yields enable investment depreciation terms to be considerably improved and plantation programmes to be drawn up which are financially feasible.

WOOD QUALITY

Another reason behind the introduction of exotic trees is the production of wood having a different quality from that of local woods. Formerly it was the production of high quality wood which attracted attention. The plantations of *Tectona grandis* in Africa and America, the spread of the various southern pines outside the southern United States, tests of the American ash *(Fraxinus americana)* and walnut *(Juglans nigra)* in Europe are well-known examples of this kind. Nowadays it is rather the great demand for coniferous wood and pulp wood that determines the type of plantations.

The plantations of *Pinus radiata* in South Africa and New Zealand, and the plantations of *Populus* species in Europe strongly reflect these tendencies.

To sum up, all four reasons mentioned here as justifying the introduction of new tree species may be collated under a single aim - that of finding the means of producing in the shortest possible time on a site the greatest quantity of wood of the desired quality for industries based on wood.

It is therefore necessary to make economic decisions based on a knowledge of the present and future local conditions of the wood market, of trends in its use, and of the prospects and possibilities of developing industries which the availability of wood might affect. The validity of these statements is confirmed by the paper industry in New Zealand based on plantations of *Pinus radiata*, and by the programme to plant *Eucalyptus* and establish a pulp industry in Morocco.

The geneticist is therefore called upon to draw up his programme of work in close co-operation with economists and to keep within the overall plan for the development of his country.

It should be remembered that there is not complete agreement about the importance of exotic trees in the forest economy. A swing of opinion against exotics has arisen in the past because of disappointments from certain introductions, too optimistic an evaluation of the possibilities of some species and, chiefly, too early a passage from the experimental phase to large-scale plantations (Edwards, 1963).

DEVELOPING A PROGRAMME OF INTRODUCTION

There are two principal kinds of problems in developing a programme to introduce exotic trees. The first concerns basic questions as regards choice of tree species, and a very effective treatment of these was recently made by Champion and Brasnett (1960) on the initiative of FAO. The greatest problem is how to obtain sufficiently detailed data on the various species, which are often little known even in their country of origin.

Preliminary knowledge of the ecology of the species in its natural range is clearly indispensable for any comparative study between the region of origin and the region in which the species is to be used. On the other hand, tests carried out on the sites in question provide the only means of assessing the productive capacity and plasticity of the species in the new environment; and these features are also linked to the variability of the species.

The introduction or the improvement of exotic species requires, therefore, basic research, and questions of genetics must occupy a very extensive place. Part of this research deals with the species in its natural range and should thus be done in the country of origin, the most important questions being the extent of genetic variability, ecological studies, and the selection of seed stands and plus trees. Introduction tests (in arboreta and test plantations) and tests of resistance and plasticity should clearly be done in the receiving country.

Secondly, there is a whole series of questions concerned with tree improvement, such as hybridization, the induction of mutations, investigations on early diagnosis in phytotrons, and so on, which, according to the individual case, may be done in the country of origin or in the host country or in both. In this connection, mention must once again be made of the importance of the exchange of information and international co-operation.

The international provenance tests on *Pinus sylvestris* (1907, 1938), *Picea abies* (1938) and *Larix decidua*; (1944); the work of Section 22 of IUFRO which deals with the study of forest plants; the results of investigations made in close co-operation by research workers from different countries, all show that teamwork should be extended on a worldwide scale.

PROBLEMS OF IMPROVEMENT IN THE COUNTRY OF ORIGIN

STUDIES ON GENETIC VARIABILITY

A close study of a species in its natural habitat can give a picture of its silvicultural and economic potential. The size to which a tree can grow, its rate of growth, and its ecological and edaphic requirements are all fundamentally the same in the country of origin and elsewhere.

Greatest attention is paid to the variability of the species which often - although not always - depends upon the extent of the area of growth. A species with a very limited range will probably not show great variability. Nevertheless, it should be remembered that the three populations which make up the natural range of *Pinus radiata* do show considerable differences; and other examples of this phenomenon are provided by *Larix leptolepis* in Japan and *Picea omorica* in Yugoslavia.

STUDIES ON GEOGRAPHIC RACES

There is always a certain amount of variation among tree species with a very extensive natural range. This variability is sometimes very marked, to the point of justifying the recognition of geographic or ecological races. The geographic variation of *Pinus sylvestris* is one of the best-known examples, and this has been elucidated by a great number of comparative experiments in Europe and North America.

Pinus ponderosa shows just as great variation and the different values of its botanical varieties and numerous races have been demonstrated by the contradictory results obtained from Europe and from the provenance tests now in progress.

Zobel (1961) has stressed the variability of *Pinus taeda* and *P. elliotii* in the southern United States and the even greater variation of *P. flexilis* and *P. montezumae* in Mexico. In this connection the biochemical research of Mirov (1962) is of great interest. In Mexico the importance and extent of this variation suggests that certain groups of species are still in the full process of evolution. Under these conditions the geneticist is able to find precious material for his work.

Certain species, particularly European and North American conifers, have already been studied in detail. A great deal has yet to be learned but, by and large, there is already sufficient basic data. In those countries where scientific forestry has just made its appearance the geneticist has an enormous field of work. What is known, for instance, of the *Swietenias* or the *Araucarias* or the *Acacias*, or other tropical species of great forest value?

For each genus, and for each species which is worth introducing, there must first be a study of the conditions in its natural range and one must distinguish, according to circumstances, the morphological, ecological and edaphic differences and examine the technical properties of the wood on the different sites.

SELECTION OF SEED STANDS AND PLUS TREES

Because of the variation shown by the great majority of tree species, the provision of seed of known origin is crucial. It is known that this problem is often difficult to resolve even in those countries with a long forestry tradition where the seed crop and its distribution have long been subjected to more or less efficient control. Germany and Sweden have always been to the fore in this respect, and several countries are following their example. In scientific work, co-operation between research institutes often results in the supply of material of well-defined origin, but sometimes, even in forest research and often in forest practice, it is not possible to get such material, especially of species which are still little known and from certain countries.

Species of the genus *Eucalyptus* provide an important example. Because of the increasing use of *Eucalyptus* the demand for Australian seed has increased very rapidly, both for reforestation and for research. It was above all the supply of seed for research, often of the rarer species or special provenances, which complicated the work. The goodwill and co-operative spirit shown by the Australian Government have led to the formation of a section of the Forestry and Timber Bureau at Canberra specially charged with the supply of seed to foreign countries.

This is a happy solution but one which unfortunately so far finds no counterpart in other countries. In fact, reliable guarantee of seed origin is no easy matter. Seed stands, and in the later phases of breeding, plus trees must be selected, often in regions which are badly served with roads and transport. Moreover, it is necessary to organize the collection of the material in the forest and keep a check on its processing right up to delivery. This problem does not solely concern exotic trees, but it is particularly accentuated in certain of these species.

In some cases the countries interested have arranged *ad hoc* expeditions to the country of origin. South Africa, New Zealand, Rhodesia and the United States (Loock, 1950; Hinds and Larsen, 1961; Zobel, 1961) sent experts to Mexico to collect seeds of Pinus species. This interest, stressed by the Seminar and Study Tour of Latin-American Conifers organized by FAO in 1960, has led the Mexican Government to organize, with the assistance of FAO, a center for the selection of seed stands of these pines, and for the collection and distribution of the seeds.

Similar seed collecting expeditions will probably be necessary to obtain seeds of other species, particularly those growing in remote areas and in countries where trained seed collectors are not available. Where collections of research material can be adequately supervised by local staff, all or part of the costs may have to be defrayed by the country or countries making the request. It is also desirable that demands for seeds that are difficult to obtain should be well thought out.

The question of seed origin is such a basic one that the authorities of every country should make it their responsibility. European foresters have not yet obtained any real guarantee, for instance, regarding the bulk importation of seed of *Pseudotsuga taxifolia* from America.

It is desirable that international co-operation, which has been greatly facilitated by the bodies which organized the World Consultation in Sweden, should lead to a solution of these problems. Members of (FAO), and IUFRO can play an important role by providing information and facilities for international seed collection and exchange. In this respect, the following are worthy of urgent attention:

1. Production and circulation of a list of research workers who are willing and able to collect seed on behalf of other countries;
2. Production of a practical code of procedure for standardizing the selection of seed trees and seed stands and site descriptions for both research and practical purposes; the importance of obtaining local meteorological data should be emphasized.

PROBLEMS OF TREE IMPROVEMENT IN TUB INTRODUCING COUNTRIES

TOLERANCE AND PLASTICITY

Although a comparative study of climatic and other ecological conditions of the region of origin and the region of introduction permits general forecasts of the chances of success, only plantation tests make possible valid conclusions about the economic value of the species, and this is the basis for any large investment.

It is unusual to obtain very detailed information about the region of origin and certain phenomena escape precise assessment. The plasticity of a species is very important because some show very marked requirements and cannot adapt themselves even to conditions which are apparently very near those which obtain in their natural range. By contrast there are other species which survive and grow vigorously in apparently very different conditions. A striking example of such differences in plasticity is given by the genus *Eucalypus* which includes several hundred species covering a more or less widespread range. Some of the species will spread only in their own particular habitat or under conditions which are strictly similar while others, even those with a very restricted range, are able to adapt themselves to very different ecological conditions.

In this context *Pinus radiata*, may again be mentioned. This species was taken from a limited area of origin and transported with great success to very different regions where the climatic variations cover an extremely wide range. Again, some species which are of no particular interest in their country of origin because of their slow and limited growth, have shown exceptional vigour as exotics.

The choice of species to introduce is more difficult when the introducing country has extreme ecological conditions. Resistance to certain factors, or adaptability to difficult habitats, are then often of greater importance than growth rate or wood characters.

Resistance to drought and cold are frequently sought after, particularly in the introduction of exotic trees to the Mediterranean region where the unreliability of the climate increases environmental difficulties. Otherwise the limiting factor in the region is that of exceptional soil conditions, particularly salinity.

Resistance to stated factors and plasticity are sometimes found throughout a species, especially in one with a very limited natural area such as *Pinus radiata*, But more frequently these characters are linked to geographical and ecological races. Attempts at introduction should take these two possibilities into account.

TESTS OF RACES AND PROVENANCES

The existence of differentiated races within a species complicates the problem. If the preliminary study enables attention to be concentrated on a more limited sector of the range and makes possible the elimination of races from other sectors, it is still necessary to consider the introduction of as wide a range of provenances as possible, and to establish tests to compare these provenances in different environments. Apart from provenances taken directly from the region of origin, the possibility should be considered of seeking material from other countries where the species has already been introduced with success. Quite often such provenances, already subjected to natural selection under the conditions of the intermediary country, have a plasticity which is far superior to that of material taken directly from the country of origin. An example of this is provided by *Eucalyptus camaldulensis.*

In some countries at the present day it is rather unusual to find that a new species is being introduced for the very first time. The presence of plantations, even small plots or strips or of trees in gardens and arboreta, is a great advantage. These provide the basis for useful preliminary studies and may represent a precious source of material.

On the basis of these plantations or individual trees which could provide a natural spread of the introduced species (which is sometimes to be found in the second or third generation as seminatural stands), it is possible to begin the work of selection while awaiting the first results of new introductions or of provenance tests which require a longer period of time and hardly ever less than ten years.

If the early introductions have reached a considerable age, they will have been subjected to climatic extremes and their state of health and growth will make it possible to draw conclusions and use them in the selection programme, particularly with regard to resistance to cold.

HYBRIDIZATION

The main aim of hybridization is to bring together in a new individual the favorable characters of the parent trees. However, the effect of increased vigour, which can occur in F_1 hybrids, often assumes greater importance. The cross may be between an exotic parent tree and a tree of an indigenous species, or between two exotic parent trees.

In the first case, there is a general tendency to improve certain characters of the indigenous species, such as wood quality, stem form, and especially resistance to disease. Thus the hybridization programmes in the United States and in Italy with the indigenous chestnuts *(Castanea dentata and C. saliva)* and the Japanese and Chinese chestnuts *(Castanea crenata and C. mollissima)* aim at making the indigenous trees resistant to ink-disease and to canker.

Hybridization between *Abies alba* and *A. veitchii* in Germany aim at improving the resistance of the indigenous silver fir to frost. Crosses between European and American poplars (such as *Populus nigra* x *deltoides)* on the other hand, stress the factor of growth rate.

Hybridization programmes using exotic species are in progress in several countries. In the United States the Institute of Forest Genetics in California places great importance on this means of improvement. In Korea, members of the Institute of Forest Genetics at Suwon, Kyunggido, are working to increase volume production.

The origin of hybrids is often fortuitous; this is the case with species whose natural ranges are separate. When individuals of each species have been brought together by chance in plantations they have given rise to hybrids which have demonstrated superior characters and have then been reproduced artificially. In *Eucalyptus*, mention can be made of the hybrids *E.* x *algeriensis, (E. camaldulensis* x *rudis x tereticornis), E. alba* x *saligna* and *E. saligna* x *grandis*, which have already achieved a well-defined place in the culture of Eucalypts. But the attention of geneticists today is directed towards the natural or artificial hybridization of many Pinus species. Many examples could be quoted here, but the most interesting include *P. strobus* x *monticola* and *P. sylvestris* x *densiflora* in America, *P. taeda* x *elliottii in Korea,* and *P. attenuata x radiata* in America and New Zealand.

SEED SUPPLY AND SEED ORCHARDS

The supply of seed mainly limits the introduction of exotic trees, especially in forming large-scale plantations. This problem has already been mentioned in this chapter, and attention is again drawn to the paragraph referring to the importation of seed from the country of origin.

Careful inspection for pests and diseases of all imported seed, cuttings and scions should be made by the importing country; adequate quarantine facilities should be available for handling cuttings and scions. It must be remembered that not all exotic species are of value, and precautions should be taken to insure that undesirable freely-seeding species do not become serious weed problems.

Old plantations are sometimes good seed sources. All early introductions should be examined and, if worthwhile, should be classified with reference to seed collection. The risks of obtaining undesirable inbred plants from seed collected from small groups and isolated trees must be carefully considered. For this reason, collections from such sources are not recommended. On the whole, the general criteria for the classification of seed stands will be found valid and the same holds for the choice of plus trees.

There is now a marked tendency to concentrate seed production in seed orchards. This practice eventually insures the supply of seed of known origin, thus avoiding the hazards involved in importation, particularly with regard to provenance. It will suffice to note the problems associated with seeds of *Pseudotsuga taxifolia, Eucalyptus* and *Pinus radiata* to appreciate the importance of this method. The basic material for the formation of these seed orchards is carefully selected on representative sites or the material is introduced and subjected to progeny tests. The production of hybrid seed poses special problems. Several countries already have their mass-production programmes, either by controlled pollination or by natural crossing, an example being the mixed plantations of *Eucalyptus viminalis* and *E. camaldulensis* for the production of hybrid seed in Morocco.

SURVEY OF SOME SPECIES

The ecological properties of each species, its production possibilities and the economic role it can play in the receiving countries are all factors influencing the prospects for the improvement of the species, the research lines to follow and the methods of work.

It is not possible to examine here the details of the improvement of every species. Five species and genera of great current interest have therefore been chosen, on which large-scale improvement work is either in progress or projected. These species are Douglas fir, the poplars, the eucalypts, Monterey pine, and the Mexican pines.

DOUGLAS FIR (PSEUDOTSUGA TAXIFOLIA BRITT)

Among the species introduced into Europe, Douglas fir takes pride of place. Experiments over several decades hare already supplied valid basic data for the adoption of this species in European forests. Indeed the plantations already cover very extensive areas in several countries and, in Europe as a whole, may be estimated at tens of thousands of hectares; the area is still increasing at a rapid rate.

The geographical and ecological extent of the natural range of *Pseudotsuga taxifolia* early set American foresters the problem of defining races and provenances. Provenance tests were established in the United States as long ago as 1911, and since then have been extended and multiplied. They have shed much light on the existence of a large number of ecological races of very varying importance to forestry. In 1949 Isaac drew up a very detailed distribution map showing Douglas fir provenance zones, accompanied by basic climatic documentation for foresters of both the countries of origin and reception.

In Europe the very variable results from the first introductions of *Pseudotsuga taxifolia* also attracted the attention of foresters. The problems of systematics and provenance have long been investigated in several countries. If productive capacity and rate of growth are the point of departure, resistance to disease, and especially to *Rhabdocline pseudotsugae*, resistance to frost, plasticity, and drought tolerance are the objects of improvement in the various countries of Europe.

It has already been suggested that for north-central Europe, provenances from the coastal zones of British Columbia and Washington give the best results, while in the southern countries of Europe provenances from Oregon and from North California show better tolerance of the dry climate. Experiments on this subject are worth extending and renewing in several countries.

The fundamental problem for European foresters is now the supply of Douglas fir seed. Seed production in America is infrequent and very often inadequate to meet the great demand, both locally and from Europe. Moreover, although some enterprises send certificates of origin with their deliveries of seed, in general there are insufficient guarantees of seed origin.

The oldest plantations of *Pseudotsuga taxifolia* in Europe are already showing good seed production, and some countries have selected stands for the production of seed. This material is of great importance because its adaptation to local climatic conditions is already known and should thus be a point of departure for genetic improvement. Several European countries have already set up Douglas fir seed orchards, and it is suggested that this is worth developing as the best means of solving the seed supply problem. The natural range of poplars covers almost the whole of the temperate zone and even extends to the subtropical zone. This extensive range, the great number of species and their phenotypic and genotypic heterogeneity demand the use of different improvement methods according to the region and the species. In the northern part of the range, poplars of section *Leuce* Duby dominate (especially *Populus tremula*). They are rather difficult to propagate vegetatively. Improvement in section *Leuce* should be based on the following points:

1. Production of selected seed by the normal methods used for forest tree seed, (that is, selection of seed stands and plus trees and control of seed collection and distribution, etc.);
2. Vegetative reproduction of stump shoots or root cuttings from selected individuals (and eventually by stem cuttings-using growth substances);
3. Production of hybrids with improved capacity for vegetative reproduction by crossing with *P. alba* among others.

Poplars of section *Tacamahaca* Spach, are also of some interest in this region.

In the warm-temperate regions, poplars of section *Aigeiros* Duby are the most important, and especially *P. nigra* L. from Europe, *P. deltoides* Marsh. from America, and their hybrids. In these poplars, vegetative propagation generally does not present difficulties. Clone formation is easy and this permits the maximum use of the hybrid vigour or heterosis shown by first-generation hybrids.

In the southern regions, where drought resistance is an important character, poplars of sections *Turanga* Bunge and *Leucoides* Spach are found alongside *P. nigra* and *P. alba* Species of the latter section present some difficulties in vegetative propagation and it is then necessary to use the improvement methods described for section *Leuce*.

A study of mutations, and particularly of natural or artificial polyploids, opens up considerable prospects for improving poplars. Researchers on poplars should consider the different ecological and economic environments, but in addition to production characters, resistance to pests and disease deserve greatest attention.

Sekawin (1963) considers that future work on the genetic improvement of poplars should be done along the following lines:

1. Precise definition of the objects of genetic improvement in relation to physical, economic and social factors;
2. A combined big-systematic study of the poplars existing in each country and of their relationship with the environment;
3. Hybridization and selection among local types;
4. International exchange of selected material and hybridization with indigenous material, together with the establishment of poplar collections;

5. Studies on the inheritance of the most important characters;
6. Research on polyploidy and mutation;
7. Improvement of the methods of reproduction and multiplication;
8. Systematic experimentation with cultivars and the clones obtained by selection and breeding.

The importance of the genus *Eucalyptus* in providing fast-growing species needs no emphasis. As exotics these species have often shown a production capacity far above that in Australia, and each year they become more and more important in the forest economy of the world (Metro, 1954, 1963), especially in the tropical and subtropical regions.

One of the most important properties of *Eucalyptus* which is especially developed in certain species is plasticity. This enables the trees to develop with great vigour in very different ecological conditions from those in their natural range. Furthermore, species of *Eucalyptus* are able to adapt themselves to new conditions, and transmit this capacity for adaptation to their progeny.

In addition, the ease of inter- and intraspecific hybridization frequently leads to the creation of new hybrids and some of these are already being used on a large scale (examples being *E.* x *algeriensis, E. alba* and *E.* 'Mysore hybrid').

On the other hand, several countries encounter difficulties in using the timber of Eucalyptus. In particular, the wood derived from rapidly grown plantations may show twisted fibres, collapse and other defects.

The main obstacle to making new introductions of *Eucalyptus* species is the difficulty of obtaining seed of known sources of the species and the provenances required. This problem has always been discussed at meetings dealing with *Eucalyptus* and especially at the second World Eucalyptus Conference at São Paulo (FAO, 1961) Selection of seed stands and the formation of seed orchards in each reception region will help to solve these difficulties. Another important method is vegetative propagation and a technique effective for several species has been already developed.

Research on the improvement of *Eucalyptus* has three objects: to extend the culture of eucalyptus; to improve production; and to improve technical properties. The investigations made to extend the use of *Eucalyptus* to new regions are based on the plasticity of certain species and provenances, the accent being placed on resistance to drought in the Mediterranean region and on resistance to cold. The selections made in France and Italy, especially in *E. gunnii, E. dalrympleana, E. rubida, E. viminalis* and *E. bridgesiana* seem to offer good opportunities of obtaining cold resistance.

The improvement of productivity is primarily based on mass and individual tree selection, and the possibility of vegetative propagation then becomes a basic consideration.

Regarding mass selection, the importance of rigorous selection of material in the nursery is stressed. This gave remarkable results at Rio Claro in Brazil. Natural and artificial hybridization and the mass production of hybrid seed also permit new types to be created and reproduced, and their hybrid vigour or heterosis exploited.

The improvement of technical properties, a subject on which Australian researchers have already achieved good results, should be based primarily on individual tree selection.

Pinus radiata provides the most striking example of the plasticity of a species. From its natural range, which is limited to three stands of small dimensions in California and to some groups of trees on the islands of Guadalupe and Cedros (Baja California, Mexico), it has been introduced with great success to regions with very different climates, such as Chile, New Zealand, South Africa and northern Spain. Monterey pine covers the largest area of all exotic tree species.

Although the natural range of Pinus radiata is so concentrated, it appears that ecological differences can be detected between the three source islands in California, and these differences are very probably far more marked at the Mexican sites of Baja California (*P. radiata* var. *binata*).

In those countries where the species has been introduced for many years and over extensive areas it is probable that, because of segregation, migration of genes and hybridization between different genotypes, the actual population has a genetic composition quite different from any other wild population (Bannister, 1963).

According to Scott (1960), there are very good stands of Monterey pine in Australia, Chile, New Zealand, South Africa and Spain with, nevertheless, a very great range of morphological characters. Forest geneticists and tree breeders in Australia and New Zealand (Bannister, 1954; Fielding, 1957; Thulin, 1957) have already made important selection work, dealing with both seed stands and plus trees. The natural hybridization of *P. radiata* with *P. attenuata* or *P. muricata*, already noted in both California and New Zealand, has also been studied. In the United States and Spain some importance is given to artificial hybridization.

It would be interesting to look further into the question of provenances in relation to plasticity, with a view to establishing plantations of Monterey pine on sites with a comparatively dry, cold climate.

MEXICAN PINES

Zobel (1961) has described Mexico as "the melting pot of the genus *Pinus*" Over a comparatively small area of Mexico there is an extensive series of Pinus species the range of variability, systematic position and forestry importance of which are far from well defined. Various authors have ranked as species a number of entities varying from 30 to 80 (Martinez, 1948). But it must be emphasized that within several species there is a very marked morphological variability which makes it possible to claim the presence of several "complexes of species," the evolution of which is far from being completed.

On the other hand, the importance of some of these pines in South Africa, especially *P. patula* (Loock, 1950), and the interest aroused in the FAO study tours in Mexico in 1960 (Morandini, 1961) have already attracted the attention of foresters in tropical and subtropical countries, who see in the Mexican pines a precious means for the rapid production of wood. The report of Hodgson gives a very important survey of the lines to follow in improving *Pinus patula*, which is today the most widespread of the Mexican pines.

This, therefore, is an ideal field of action by specialists in the various branches of genetics. Without doubt, the point of departure is a systematic study of the different units and an analysis of their morphogenetic variation. Ecological studies will enable the researcher to determine the existence and the extent of ecological races, which in some species with an extensive range will no doubt show very marked differences. In this respect, it is necessary to stress the importance of races comparatively tolerant to cold, because they would permit the use of some species in regions with a less mild climate (such as the Mediterranean area).

The selection of seed stands and plus trees will enable seed production to be organized on a rational basis, because the demand for seed is already quite sizable. In this context, a prominent part can be played by the technical assistance programmes of FAO; a highly-qualified researcher has already been sent to Mexico to cooperate with Mexican foresters in organizing seed supplies. While the fulcrum for the improvement of these pines will remain their country of origin, it must be stressed that several countries, especially South Africa, have already launched programmes of improvement based on their existing plantations, some of which are more than 50 years old.

CONCLUSIONS

A programme for the introduction of exotic trees and their improvement should be based on the following points:

INTRODUCTION OF THE SPECIES

1. Studies of climatic similarities between the various regions of the world; the *Bio-climatic map of the Mediterranean zone* recently published by Unesco and FAO (1963) provides an excellent example.
2. Biological and economic analysis of the forest flora of similar climatic regions, and the choice of quick-growing species, the wood of which

is readily utilizable in large-scale industries such as paper manufacture.

3. Ecological studies to distinguish the most suitable races and provenances.
4. Tests to screen out, by planting small groups in arboreta, a large number of species and test plantations (on sufficiently large areas to insure the statistical value of the results) of those species which have good prospects of success.

IMPROVEMENT OF THE SPECIES

1. Provenance tests.
2. Selection for both quantitative and qualitative production.
3. Large-scale production of seed and cuttings for selection in seed stands and seed orchards.

It is clear that such a programme will take a long time and require sizable investments. Hence it will be useful to concentrate on a fairly restricted number of species whose products have an assured economic value. Finally, it is important to refer once more to the importance of international co-operation, which makes it possible to turn to good account the experience already acquired, and to concentrate efforts on achieving further progress.

6

Carbon Sink

INTRODUCTION

A carbon sink is a natural or artificial reservoir that accumulates and stores some carbon-containing chemical compound for an indefinite period. The process by which carbon sinks remove carbon dioxide from the atmosphere is known as carbon sequestration. Public awareness of the significance of CO_2 sinks has grown since passage of the Kyoto Protocol, which promotes their use as a form of carbon offset. The main natural sinks are:

- Absorption of carbon dioxide by the oceans via physicochemical and biological processes
- Photosynthesis by terrestrial plants

Natural sinks are typically much larger than artificial sinks. The main artificial sinks are:

- Landfills
- Carbon capture and storage proposals

Carbon sources include:

- Fossil fuels
- Farmland; there are proposals for improvements in farming practices to reverse this.

KYOTO PROTOCOL

Because growing vegetation absorbs carbon dioxide, the Kyoto Protocol allows Annex I countries with large areas of growing forests to issue Removal Units to recognise the sequestration of carbon. The additional units make it easier for them to achieve their target emission levels. Some countries seek to trade emission rights in carbon emission markets, purchasing the unused carbon emission allowances of other countries. If overall limits on greenhouse gas emission are put into place, cap and trade market mechanisms are purported to find cost-effective ways to reduce emissions. There is as yet no carbon audit regime for all such markets globally, and none is specified in the Kyoto Protocol. National carbon emissions are self-declared. In the Clean Development Mechanism, only afforestation and reforestation are eligible to produce certified emission reductions in the first commitment period of the Kyoto Protocol. Forest conservation activities or activities avoiding deforestation, which would result in emission reduction through the conservation of existing carbon stocks, are not eligible at this time. Also, agricultural carbon sequestration is not possible yet.

STORAGE IN TERRESTRIAL AND MARINE ENVIRONMENTS

SOILS

Soils represent a short to long-term carbon storage medium, and contain more carbon than all terrestrial vegetation and the atmosphere combined. Plant litter and other biomass accumulates as organic matter in soils, and is degraded by chemical weathering and biological degradation. More recalcitrant organic carbon polymers such as cellulose, hemi-cellulose, lignin, aliphatic compounds, waxes and terpenoids are collectively retained as humus. Organic matter tends to accumulate in litter and soils of colder regions such as the boreal forests of North America and the Taiga of Russia. Leaf litter and humus are rapidly oxidized and poorly retained in sub-tropical and tropical climate conditions due to high temperatures and extensive leaching by rainfall.

Areas where shifting cultivation or slash and burn agriculture are practiced are generally only fertile for 2–3 years before they are abandoned. These tropical jungles are similar to coral reefs in that they are highly efficient at conserving and circulating necessary nutrients, which explains their lushness in a nutrient desert. Much organic carbon retained in many agricultural areas worldwide has been severely depleted due to intensive farming practices. Grasslands contribute to soil organic matter, stored mainly in their extensive fibrous root mats.

Due in part to the climactic conditions of these regions, these soils can accumulate significant quantities of organic matter. This can vary based on rainfall, the length of the winter season, and the frequency of naturally occurring lightning-induced grass-fires. While these fires release carbon dioxide, they improve the quality of the grasslands overall, in turn increasing the amount of carbon retained in the retained humic material. They also deposit carbon directly to the soil in the form of char that does not significantly degrade back to carbon dioxide.

Forest fires release absorbed carbon back into the atmosphere, as does deforestation due to rapidly increased oxidation of soil organic matter. Organic matter in peat bogs undergoes slow anaerobic decomposition below the surface. This process is slow enough that in many cases the bog grows rapidly and fixes more carbon from the atmosphere than is released. Over time, the peat grows deeper. Peat bogs inter approximately one-quarter of the carbon stored in land plants and soils. Under some conditions, forests and peat bogs may become sources of CO_2, such as when a forest is flooded by the construction of a hydroelectric dam. Unless the forests and peat are harvested before flooding, the rotting vegetation is a source of CO_2 and methane comparable in magnitude to the amount of carbon released by a fossil-fuel powered plant of equivalent power.

REGENERATIVE AGRICULTURE

Current agricultural practices lead to carbon loss from soils. It has been suggested that improved farming practices could return the soils to being a carbon sink. The Rodale Institute says that Regenerative agriculture, if practiced on the planet's 3.5 billion tillable acres, could sequester up to 40 per cent of current CO_2 emissions. They claim that agricultural carbon sequestration has the potential to mitigate global warming. When using biologically based regenerative practices, this dramatic benefit can be accomplished with no decrease in yields or farmer profits. Organically managed soils can convert carbon dioxide from a greenhouse gas into a food-producing asset.

In 2006, U.S. carbon dioxide emissions from fossil fuel combustion were estimated at nearly 6.5 billion tons. If a 2,000 year sequestration rate was achieved on all 434,000,000 acres of cropland in the United States, nearly 1.6 billion tons of carbon dioxide would be sequestered per year, mitigating close to one quarter of the country's total fossil fuel emissions.

OCEANS

Oceans are at present CO_2 sinks, and represent the largest active carbon sink on Earth, absorbing more than a quarter of the carbon dioxide that humans put into the air. On longer timescales they may be both sources and sinks - during ice ages CO_2 levels decrease to ~180 ppmv, and much of this is believed to be stored in the oceans. As ice ages end, CO_2 is released from the oceans and CO_2 levels during previous interglacials have been around ~280 ppmv. This role as a sink for CO_2 is driven by two processes, the solubility pump and the biological pump.

The former is primarily a function of differential CO_2 solubility in seawater and the thermohaline circulation, while the latter is the sum of a series of biological processes that transport carbon from the surface euphotic zone to the ocean's interior. A small fraction of the organic carbon transported by the biological pump to the seafloor is buried in anoxic conditions under sediments and ultimately forms fossil fuels such as oil and natural gas. At the present time, approximately one third of human generated emissions are estimated to be entering the ocean. The solubility pump is the primary mechanism driving this, with the biological pump playing a negligible role. This stems from the limitation of the biological pump by ambient light and nutrients required by the phytoplankton that ultimately drive it. Total inorganic carbon is not believed to limit primary production in the oceans, so its increasing availability in the ocean does not directly affect production. However, ocean acidification by invading anthropogenic CO_2 may affect the biological pump by negatively impacting calcifying organisms such as coccolithophores, foraminiferans and pteropods. Climate change may also affect the biological pump in the future by warming and stratifying the surface ocean, thus reducing the supply of limiting nutrients to surface waters. In January 2009, the Monterey Bay Aquarium Research Institute and the National Oceanic and Atmospheric Administration announced a joint study to determine whether the ocean off the California coast was serving as a carbon source or a carbon sink. Principal instrumentation for the study will be self-contained CO_2 monitors placed on buoys in the ocean. They will measure the partial pressure of CO_2 in the ocean and the atmosphere just above the water surface. In February 2009, Science Daily reported that the Southern Indian Ocean is becoming less effective at absorbing carbon dioxide due to changes to the regions climate which include higher wind speeds.

ENHANCING NATURAL SEQUESTRATION

FORESTS

Forests are carbon stores, and they are carbon dioxide sinks when they are increasing in density or area. In Canada's boreal forests as much as 80 per cent of the total carbon is stored in the soils as dead organic matter. A 40-year study of African, Asian, and South American tropical forests by the University of Leeds, shows tropical forests absorb about 18 per cent of all carbon dioxide added by fossil fuels. Tropical reforestation can mitigate global warming until all available land has been reforested with mature forests. However, the global cooling effect of carbon sequestration by forests is partially counterbalanced in that reforestation can decrease the reflection of sunlight. Mid-to-high latitude forests have a much lower albedo during snow seasons than flat ground, thus contributing to warming. Modeling that compares the effects of albedo differences between forests and grasslands suggests that expanding the land area of forests in temperate zones offers only a temporary cooling benefit. In the United States in 2004, forests sequestered 10.6 per cent of the carbon dioxide released in the United States by the combustion of fossil fuels. Urban trees sequestered another 1.5 per cent. To further reduce U.S. carbon dioxide emissions by 7 per cent, as stipulated by the Kyoto Protocol, would require the planting of "an area the size of Texas every 30 years". Carbon offset programmes are planting millions of fast-growing trees per year to reforest tropical lands, for as little as $0.10 per tree; over their typical 40-year lifetime, one million of these trees will fix 0.9 teragrams of carbon dioxide. In Canada, reducing timber harvesting would have very little impact on carbon dioxide emissions because of the combination of harvest and stored carbon in manufactured wood products along with the regrowth of the harvested forests.

Additionally, the amount of carbon released from harvesting is small compared to the amount of carbon lost each year to forest fires and other natural disturbances. The Intergovernmental Panel on Climate Change concluded that "a sustainable forest management strategy aimed at maintaining or increasing forest carbon stocks, while producing an annual sustained yield of timber fibre or energy from the forest, will generate the largest sustained mitigation benefit". Sustainable management practices keep forests growing at a higher rate over a potentially longer period of time, thus providing net sequestration benefits in addition to those of unmanaged forests. Life expectancy of forests varies throughout the world, influenced by tree species, site conditions and natural disturbance patterns. In some forests carbon may be stored for centuries, while in other forests carbon is released with frequent stand replacing fires. Forests that are harvested prior to stand replacing events allow for the retention of carbon in manufactured forest products such as lumber. However, only a portion of the carbon removed from logged forests ends up as durable goods and buildings. The remainder ends up as sawmill by-products such as pulp, paper and pallets, which often end with incineration at the end of their lifecycle. For instance, of the 1,692 teragrams of carbon harvested from forests in Oregon and Washington from 1900 to 1992, only 23 per cent is in long-term storage in forest products.

OCEANS

One way to increase the carbon sequestration efficiency of the oceans is to add micrometre-sized iron particles in the form of either hematite or melanterite to certain regions of the ocean. This has the effect of stimulating growth of plankton. Iron is an important nutrient for phytoplankton, usually made available via upwelling along the continental shelves, inflows from rivers and streams, as well as deposition of dust suspended in the atmosphere. Natural sources of ocean iron have been declining in recent decades, contributing to an overall decline in ocean productivity. Yet in the presence of iron nutrients plankton populations quickly grow, or 'bloom', expanding the base of biomass productivity throughout the region and removing significant quantities of CO_2 from the atmosphere via photosynthesis. A test in 2002 in the Southern Ocean around Antarctica suggests that between 10,000 and 100,000 carbon atoms are sunk for each iron atom added to the water. More recent work in Germany suggests that any biomass carbon in the oceans, whether exported to depth or recycled in the euphotic zone, represents long-term storage of carbon. This means that application of iron nutrients in select parts of the oceans, at appropriate scales, could have the combined effect of restoring ocean productivity while at the same time mitigating the effects of human caused emissions of carbon dioxide to the atmosphere.

Because the effect of periodic small scale phytoplankton blooms on ocean ecosystems is unclear, more studies would be helpful. Phytoplanktons have a complex effect on cloud formation via the release of substances such as dimethyl sulfide that are converted to sulfate aerosols in the atmosphere, providing cloud condensation nuclei, or CCN. But the effect of small scale plankton blooms on overall DMS production is unknown. Other nutrients such as nitrates, phosphates, and silica as well as iron may cause ocean fertilization. There has been some speculation that using pulses of fertilization may be more effective at getting carbon to ocean floor than sustained fertilization. There is some controversy over seeding the oceans with iron however, due to the potential for increased toxic phytoplankton growth, declining water quality due to overgrowth, and increasing anoxia in areas harming other sea-life such as zooplankton, fish, coral, etc.

SOILS

Since the 1850s, a large proportion of the world's grasslands have been tilled and converted to croplands, allowing the rapid oxidation of large quantities of soil organic carbon. However, in the United States in 2004, agricultural soils including pasture land sequestered 0.8 per cent as much carbon as was released in the United States by the combustion of fossil fuels. The annual amount of this sequestration has been gradually increasing since 1998. Methods that significantly enhance carbon sequestration in soil include no-till farming, residue mulching, cover cropping, and crop rotation, all of which are more widely used in organic farming than in conventional farming. Because only 5 per cent of US farmland currently uses no-till and residue mulching, there is a large potential for carbon sequestration. Conversion to pastureland, particularly with good management of grazing, can sequester even more carbon in the soil. Terra preta, an anthropogenic, high-carbon soil, is also being investigated as a sequestration mechanism. By pyrolysing biomass, about half of its carbon can be reduced to charcoal, which can persist in the soil for centuries, and makes a useful soil amendment, especially in tropical soils.

SAVANNA

Controlled burns on far north Australian savannas can result in an overall carbon sink. One working example is the West Arnhem Fire Management

Agreement, started to bring "strategic fire management across 28,000 km^2 of Western Arnhem Land". Deliberately starting controlled burns early in the dry season results in a mosaic of burnt and unburnt country which reduces the area of burning compared with stronger, late dry season fires. In the early dry season there are higher moisture levels, cooler temperatures, and lighter wind than later in the dry season; fires tend to go out overnight. Early controlled burns also results in a smaller proportion of the grass and tree biomass being burnt. Emission reductions of 256,000 tonnes of CO_2 have been made as of 2007.

ARTIFICIAL SEQUESTRATION

For carbon to be sequestered artificially it must first be captured, or it must be significantly delayed or prevented from being re-released into the atmosphere from an existing carbon-rich material, by being incorporated into an enduring usage. Thereafter it can be passively stored or remain productively utilized over time in a variety of ways. For example, upon harvesting, wood can be immediately burned or otherwise serve as a fuel, returning its carbon to the atmosphere, or it can be incorporated into construction or a range of other durable products, thus sequestering its carbon over years or even centuries. One ton of dry wood is equivalent to 1.8 tons of carbon dioxide.

Indeed, a very carefully designed and durable, energy-efficient and energy-capturing building has the potential to sequester as much as or more carbon than was released by the acquisition and incorporation of all its materials and than will be released by building-function "energy-imports" during the structure's existence. Such a structure might be termed "carbon neutral" or even "carbon negative". Building construction and operation are estimated to contribute nearly half of the annual human-caused carbon additions to the atmosphere.

Natural-gas purification plants often already have to remove carbon dioxide, either to avoid dry ice clogging gas tankers or to prevent carbon-dioxide concentrations exceeding the 3 per cent maximum permitted on the natural-gas distribution grid.

Beyond this, one of the most likely early applications of carbon capture is the capture of carbon dioxide from flue gases at power stations. A typical new 1000 MW coal-fired power station produces around 6 million tons of carbon dioxide annually. Adding carbon capture to existing plants can add significantly to the costs of energy production; scrubbing costs aside, a 1000 MW coal plant will require the storage of about 50 million barrels of carbon dioxide a year. However, scrubbing is relatively affordable when added to new plants based on coal gasification technology, where it is estimated to raise energy costs for households in the United States using only coal-fired electricity sources from 10 cents per kW·h to 12 cents.

CARBON CAPTURE

Currently, capture of carbon dioxide is performed on a large scale by absorption of carbon dioxide onto various amine-based solvents. Other techniques are currently being investigated, such as pressure swing adsorption, temperature swing adsorption, gas separation membranes, and cryogenics. Recent pilot studies include flue capture and conversion to baking soda and use of algae for conversion to fuel or feed. In coal-fired power stations, the main alternatives to retrofitting amine-based absorbers to existing power stations are two new technologies: coal gasification combined-cycle and oxy-fuel combustion.

Gasification first produces a "syngas" primarily of hydrogen and carbon monoxide, which is burned, with carbon dioxide filtered from the flue gas. Oxy-fuel combustion burns the coal in oxygen instead of air, producing only carbon dioxide and water vapour, which are relatively easily separated. Some of the combustion products must be returned to the combustion chamber, either before or after separation, otherwise the temperatures would be too high for the turbine. Another long-term option is carbon capture directly from the air using hydroxides. The air would literally be scrubbed of its CO_2 content. This idea offers an alternative to non-carbon-based fuels for the transportation sector. Examples of carbon sequestration at coal plants include converting carbon from smokestacks into baking soda, and algae-based carbon capture, circumventing storage by converting algae into fuel or feed.

OCEANS

Another proposed form of carbon sequestration in the ocean is direct injection. In this method, carbon dioxide is pumped directly into the water at depth, and expected to form "lakes" of liquid CO_2 at the bottom. Experiments carried out in moderate to deep waters indicate that the liquid CO_2 reacts to form solid CO_2 clathrate hydrates, which gradually dissolve in the surrounding waters. This method, too, has potentially dangerous environmental consequences.

The carbon dioxide does react with the water to form carbonic acid, H_2CO_3; however, most remains as dissolved molecular CO_2. The equilibrium would no doubt be quite different under the high pressure conditions in the deep ocean. In addition, if deep-sea bacterial methanogens that reduce carbon dioxide were to encounter the carbon dioxide sinks, levels of methane gas may increase, leading to the generation of an even worse greenhouse gas. The resulting environmental effects on benthic life forms of the bathypelagic, abyssopelagic and hadopelagic zones are unknown. Even though life appears to be rather sparse in the deep ocean basins, energy and chemical effects in these deep basins could have far-reaching implications. Much more work is needed here to define the extent of the potential problems. Carbon storage in or under oceans may not be compatible with the Convention on the Prevention of Marine Pollution by Dumping of Wastes and Other Matter. An additional method of long-term ocean-based sequestration is to gather crop residue such as corn stalks or excess hay into large weighted bales of biomass and deposit it in the alluvial fan areas of the deep ocean basin. Dropping these residues in alluvial fans would cause the residues to be quickly buried in silt on the sea floor, sequestering the biomass for very long time spans.

Alluvial fans exist in all of the world's oceans and seas where river deltas fall off the edge of the continental shelf such as the Mississippi alluvial fan in the gulf of Mexico and the Nile alluvial fan in the Mediterranean Sea. A downside, however, would be an increase in aerobic bacteria growth due to the introduction of biomass, leading to more competition for oxygen resources in the deep sea, similar to the oxygen minimum zone.

GEOLOGICAL SEQUESTRATION

The method of geo-sequestration or geological storage involves injecting carbon dioxide directly into underground geological formations. Declining oil fields, saline aquifers, and unminable coal seams have been suggested as storage sites. Caverns and old mines that are commonly used to store natural gas are not considered, because of a lack of storage safety. CO_2 has been injected into declining oil fields for more than 40 years, to increase oil recovery. This option is attractive because the storage costs are offset by the sale of additional oil that is recovered.

Typically, 10-15 per cent additional recovery of the original oil in place is possible. Further benefits are the existing infrastructure and the geophysical and geological information about the oil field that is available from the oil exploration. Another benefit of injecting CO_2 into Oil fields is that CO_2 is soluble in oil. Dissolving CO_2 in oil lowers the viscosity of the oil and reduces its interfacial tension which increases the oils mobility. All oil fields have a geological barrier preventing upward migration of oil. As most oil and gas has been in place for millions to tens of millions of years, depleted oil and gas reservoirs can contain carbon dioxide for millennia. Identified possible problems are the many 'leak' opportunities provided by old oil wells, the need for high injection pressures and acidification which can damage the geological barrier.

Other disadvantages of old oil fields are their limited geographic distribution and depths, which require high injection pressures for sequestration. Below a depth of about 1000 m, carbon dioxide is injected as a supercritical fluid, a material with the density of a liquid, but the viscosity and diffusivity of a gas. Unminable coal seams can be used to store CO_2, because CO_2 absorbs to the coal surface, ensuring safe long-term storage.

In the process it releases methane that was previously adsorbed to the coal surface and that may be recovered. Again the sale of the methane can be used to offset the cost of the CO_2 storage. Release or burning of methane would of course at least partially offset the obtained sequestration result – except when the gas is allowed to escape into the atmosphere in significant quantities: methane has a higher global warming potential than CO_2. Saline aquifers contain highly mineralized brines and have so far been considered of no benefit to humans except in a few cases where they have been used for the storage of chemical waste.

Their advantages include a large potential storage volume and relatively common occurrence reducing the distance over which CO_2 has to be transported. The major disadvantage of saline aquifers is that relatively little is known about them compared to oil fields. Another disadvantage of saline aquifers is that as the salinity of the water increases, less CO_2 can be dissolved into aqueous solution. To keep the cost of storage acceptable the geophysical exploration may be limited, resulting in larger uncertainty about the structure of a given aquifer. Unlike storage in oil fields or coal beds, no side product will offset the storage cost.

Leakage of CO_2 back into the atmosphere may be a problem in saline-aquifer storage. However, current research shows that several trapping mechanisms immobilize the CO_2 underground, reducing the risk of leakage. A major research project examining the geological sequestration of carbon dioxide is currently being performed at an oil field at Weyburn in south-eastern Saskatchewan. In the North Sea, Norway's Statoil natural-gas platform Sleipner strips carbon dioxide out of the natural gas with amine solvents and disposes of this carbon dioxide by geological sequestration. Sleipner reduces emissions of carbon dioxide by approximately one million tonnes a year. The cost of geological sequestration is minor relative to the overall running costs.

As of April 2005, BP is considering a trial of large-scale sequestration of carbon dioxide stripped from power plant emissions in the Miller oilfield as its reserves are depleted. In October 2007, the Bureau of Economic Geology at The University of Texas at Austin received a 10-year, $38 million subcontract to conduct the first intensively monitored, long-term project in the United States studying the feasibility of injecting a large volume of CO_2 for underground storage. The project is a research programme of the Southeast Regional Carbon Sequestration Partnership, funded by the National Energy Technology Laboratory of the U.S. Department of Energy. The SECARB partnership will demonstrate CO_2 injection rate and storage capacity in the Tuscaloosa-Woodbine geologic system that stretches from Texas to Florida. Beginning in fall 2007, the project will inject CO_2 at the rate of one million tons per year, for up to 1.5 years, into brine up to 10,000 feet below the land surface near the Cranfield oil field about 15 miles east of Natchez, Mississippi. Experimental equipment will measure the ability of the subsurface to accept and retain CO_2.

MINERAL SEQUESTRATION

Mineral sequestration aims to trap carbon in the form of solid carbonate salts. This process occurs slowly in nature and is responsible for the deposition and accumulation of limestone over geologic time. Carbonic acid in groundwater slowly reacts with complex silicates to dissolve calcium, magnesium, alkalis and silica and leave a residue of clay minerals. The dissolved calcium and magnesium react with bicarbonate to precipitate calcium and magnesium carbonates, a process that organisms use to make shells. When the organisms die, their shells are deposited as sediment and eventually turn into limestone. Limestones have accumulated over billions of years of geologic time and contain much of Earth's carbon. Ongoing research aims to speed up similar reactions involving alkali carbonates. One proposed reaction is that of the olivine-rich rock dunite, or its hydrated equivalent serpentinite with carbon dioxide to form the carbonate mineral magnesite, plus silica and iron oxide. Serpentinite sequestration is favored because of the non-toxic and stable nature of magnesium carbonate. The ideal reactions involve the magnesium endmember components of the olivine or serpentine, the latter derived from earlier olivine by hydration and silicification. The presence of iron in the olivine or serpentine reduces the efficiency of sequestration, since the iron components of these minerals break down to iron oxide and silica.

7

Implications of Climate Change

To better understand the science assessed in this Fourth Assessment Report, it is helpful to review the long historical perspective that has led to the current state of climate change knowledge. This stage starts by describing the fundamental nature of earth science. It then describes the history of climate change science using a wide-ranging subset of examples, and ends with a history of the IPCC. There is no counterpart in previous IPCC assessment reports for an introductory stage providing historical context for the remainder of the report. Here, a restricted set of topics has been selected to show key accomplishments and challenges in climate change science. The topics have been chosen for their significance to the IPCC task of assessing information relevant for understanding the risks of human-induced climate change, and also to show the complex and uneven pace of scientific progress. The time frame under consideration stops with the publication of the Third Assessment Report. Developments subsequent to the TAR are described in the other stages of this report, and we refer to these stages throughout this first stage.

CAUSES OF CLIMATE CHANGE

The general state of the Earth's climate is dependent upon the amount of energy stored by the climate system, and in particular the balance between the amount of energy the Earth receives from the Sun, in the form of light and ultraviolet radiation, and the amount of energy the Earth releases back to space, in the form of infrared heat energy. Causes of climate change involve any process that can alter this global energy balance. Scientists call this"climate forcing". Climate forcing"forces" the climate to change.

The are many climate forcing processes, but broadly speaking, they can be separated into internal and external types. External processes operate outside planet Earth, and includes changes in the global energy balance due to variations in the Earth's orbit around the Sun, and changes in the amount of energy received from the Sun. Internal processes operates from within the Earth's climate system, and include changes in the global energy balance due to changes in ocean circulation or changes in the composition of the atmosphere. Other climate forcing processes include the impacts of large volcanic eruptions and collisions with comets or meteorites. Luckily, the Earth is not hit by large comets or meteorites very often, perhaps every 20 to 30 million years or so, and therefore their associated climate changes occur rarely throughout Earth Antiquity. However, other causes of climate change influence the Earth on much shorter time scales, with changes sometimes occurring within a single generation. Indeed, our present pollution of the atmosphere with greenhouse gases may be causing the global climate to change. This man-made climate change has become known as global warming.

CLIMATE FORCING

The Earth's climate changes when the amount of energy stored by the climate system is varied. The most significant changes occur when the global energy balance between incoming energy from the Sun and outgoing heat from the Earth is upset. There are a number of natural mechanisms that can upset this balance, for example fluctuations in the Earth's orbit, variations in ocean circulation and changes in the composition of the Earth's atmosphere. In recent times, the latter has been evident as a consequence not of natural processes but of man-made pollution, through emissions of greenhouse gases. By altering the global energy balance, such mechanisms"force" the climate to change. Consequently, scientists call them"climate forcing" mechanisms.

CLIMATE SYSTEM

The key to understanding global climate change is to first understand what global climate is, and how it operates. At the planetary scale, the global climate is regulated by how much energy the Earth receives from the Sun. However, the global climate is also affected by other flows of energy which take place within the climate system itself. This global climate system is made up of the atmosphere, the oceans, the ice sheets (cryosphere), living organisms (biosphere) and the soils, sediments and rocks (geosphere), which all affect, to a greater or less extent, the movement of heat around the Earth's surface.

The atmosphere plays a crucial role in the regulation of Earth's climate. The atmosphere is a mixture of different gases and aerosols (suspended liquid and solid particles) collectively known as air. Air consists mostly of nitrogen (78 per cent) and oxygen (21 per cent). However, despite their relative scarcity, the so-called greenhouse gases, including carbon dioxide and methane, have a dramatic effect on the amount of energy that is stored within the atmosphere, and consequently the Earth's climate. These greenhouse gases trap heat within the lower atmosphere that is trying to escape to space, and in doing so, make the surface of the Earth hotter. This heat trapping is called the natural greenhouse effect, and keeps the Earth 33°C warmer than it would otherwise be. In the last 200 years, man-made emissions of greenhouse gases have enhanced the natural greenhouse effect, which may be causing global warming. The atmosphere however, does not operate as an isolated system. Flows of energy take place between the atmosphere and the other parts of the climate system, most significantly the world's oceans. For example, ocean currents move heat from warm equatorial latitudes to colder polar latitudes. Heat is also transferred via moisture. Water evaporating from the surface of the oceans stores heat which is subsequently released when the vapour condenses to form clouds and rain. The significance of the oceans is that they store a much greater quantity of heat than the atmosphere.

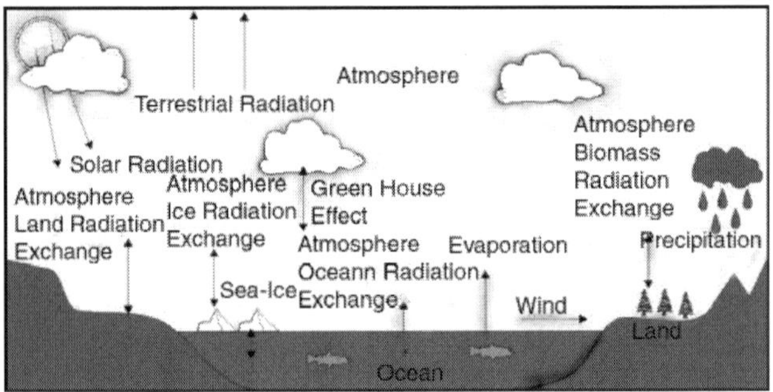

Fig. The Earth's Climate System Responds to Changes not just in the Atmosphere but also the Oceans and the Ice Sheets, and over Longer Periods of time, Movements of the Earth's crust and even the Evolution of life itself.

The top 200 metres of the world's oceans store 30 times as much heat as the atmosphere. Therefore, flows of energy between the oceans and the atmosphere can have dramatic effects on the global climate. The world's ice sheets, glaciers and sea ice, collectively known as the cryosphere, have a significant impact on the Earth's climate. The cryosphere includes Antarctica, the Arctic Ocean, Greenland, Northern Canada, Northern Siberia and most of the high mountain ranges throughout the world, where sub-zero temperatures persist throughout the year. Snow and ice, being white, reflect a lot of sunlight, instead of absorbing it. Without the cryosphere, more energy would be absorbed at the Earth's surface rather than reflected, and consequently the temperature of the atmosphere would be much higher. All land plants make food from the photosynthesis of carbon dioxide and water in the presence of sunlight. Through this utilisation of carbon dioxide in the atmosphere, plants have the ability to regulate the global climate. In the oceans, microscopic plankton utilise carbon dioxide dissolved in seawater for photosynthesis and the manufacture of their tiny carbonate shells. The oceans replace the utilised carbon dioxide by"sucking" down the gas from the atmosphere. When the plankton die, their carbonate shells sink to the seafloor, effectively locking away the carbon dioxide from the atmosphere. Such a"biological pump" reduces by at least four-fold the atmospheric concentration of carbon dioxide, significantly weakening the Earth's natural greenhouse effect, and reducing the Earth's surface temperature.

COMETS AND METEORITES

The Earth's atmosphere protects us from the impacts of comets and meteorites, by vaporising all or most of the incoming material before reaching the Earth's surface. Scientists believe however, that every once in a while a celestial body of significant size collides with the Earth, causing untold destruction and an ensuing global climate change. The comet/meteorite impact theory of climate change has been considered to account for the extinction of the dinosaurs 65 million years ago. Although critics of this theory suggest volcanism offers a better explanation for the mass extinction, evidence does exist which seems to support the case for a meteoritic impact which occurred at this time in Earth Antiquity. A proposed impact site that has been dated at 65 millions years can be found on the tip of the Yucatan Peninsula in the Gulf of Mexico, Central America.

In addition, a thin layer of clay deposited at this time at many localities around the Earth contains the rare element iridium. Although released in volcanic lava originating from deep within the Earth's interior, the impact theory indicates that the required iridium could come from a celestial object 10km across. An impact from a body this size would have released huge quantities of vaporised material into the atmosphere, blocking out the Sun and causing an initial"impact winter". Debris containing the iridium would gradually return to Earth forming the clay deposit. After this there would have been an increase in temperatures caused by the large amounts of carbon dioxide released by a spread of global fires. In addition, chemical reactions taking place in the atmosphere between pollutants would result in the formation of globally distributed acid rains.

Luckily such impacts only occur rarely, perhaps every few million or tens of millions of years. Although smaller objects hit the Earth more frequently, they have much less impact. Nevertheless, it is almost certain that another large comet or meteorite will at some time in the future strike the planet with potential consequences for the global climate and for life on Earth.

NATURE OF EARTH SCIENCE IN CLIMATE CHANGE

Science may be stimulated by argument and debate, but it generally advances through formulating hypotheses clearly and testing them objectively. This testing is the key to science. In fact, one philosopher of science insisted that to be genuinely scientific, a statement must be susceptible to testing that could potentially show it to be false. In practice, contemporary scientists usually submit their research findings to the scrutiny of their peers, which includes disclosing the methods that they use, so their results can be checked through replication by other scientists.

The insights and research results of individual scientists, even scientists of unquestioned genius, are thus confirmed or rejected in the peer-reviewed literature by the combined efforts of many other scientists. It is not the belief or opinion of the scientists that is important, but rather the results of this testing. However, that one opposing scientist would have needed proof in the form of testable results. Thus science is inherently self-correcting; incorrect or incomplete scientific concepts ultimately do not survive repeated testing against observations of nature. Scientific theories are ways of explaining phenomena and providing insights that can be evaluated by comparison with physical reality.

Each successful prediction adds to the weight of evidence supporting the theory, and any unsuccessful prediction demonstrates that the underlying theory is imperfect and requires improvement or abandonment. Sometimes, only certain kinds of questions tend to be asked about a scientific phenomenon until contradictions build to a point where a sudden change of paradigm takes place. At that point, an entire field can be rapidly reconstructed under the new paradigm.

Despite occasional major paradigm shifts, the majority of scientific insights, even unexpected insights, tend to emerge incrementally as a result of repeated attempts to test hypotheses as thoroughly as possible. Therefore, because almost every new advance is based on the research and understanding that has gone before, science is cumulative, with useful features retained and non-useful features abandoned. Active research scientists, throughout their careers, typically spend large fractions of their working time studying in depth what other scientists have done. Superficial or amateurish acquaintance with the current state of a scientific research topic is an obstacle to a scientist's progress. Working scientists know that a day in the library can save a year in the laboratory. Even Sir Isaac Newton wrote that if he had 'seen further it is by standing on the shoulders of giants'. Intellectual honesty and professional ethics call for scientists to acknowledge the work of predecessors and colleagues.

The attributes of science briefly described here can be used in assessing competing assertions about climate change. Can the statement under consideration, in principle, be proven false? Has it been rigourously tested? Did it appear in the peer-reviewed literature? Did it build on the existing research record where appropriate? If the answer to any of these questions is no, then less credence should be given to the assertion until it is tested and independently verified.

The IPCC assesses the scientific literature to create a report based on the best available science. It must be acknowledged, however, that the IPCC also contributes to science by identifying the key uncertainties and by stimulating and coordinating targeted research to answer important climate change questions. A characteristic of Earth sciences is that Earth scientists are unable to perform controlled experiments on the planet as a whole and then observe the results. In this sense, Earth science is similar to the disciplines of astronomy and cosmology that cannot conduct experiments on galaxies or the cosmos.

This is an important consideration, because it is precisely such whole-Earth, system-scale experiments, incorporating the full complexity of interacting processes and feedbacks, that might ideally be required to fully verify or falsify climate change hypotheses.

Nevertheless, countless empirical tests of numerous different hypotheses have built up a massive body of Earth science knowledge. This repeated testing has refined the understanding of numerous aspects of the climate system, from deep oceanic circulation to stratospheric chemistry. Sometimes a combination of observations and models can be used to test planetary-scale hypotheses. For example, the global cooling and drying of the atmosphere observed after the eruption of Mt. Pinatubo provided key tests of particular aspects of global climate models.

Another example is provided by past IPCC projections of future climate change compared to current observations. Figure reveals that the model projections of global average temperature from the First Assessment Report were higher than those from the Second Assessment Report.

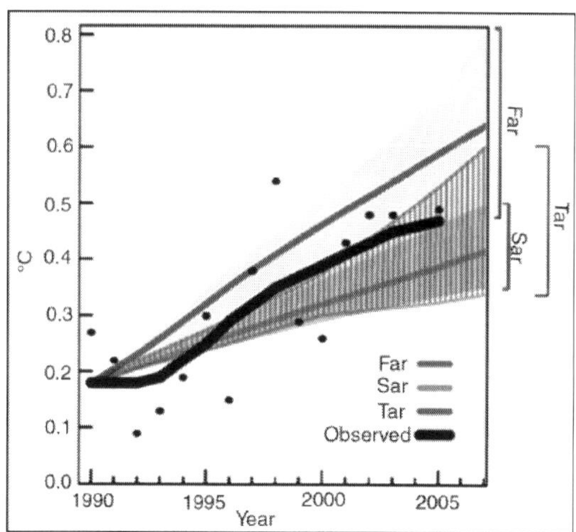

Fig. Yearly Global Average Surface Temperature

Subsequent observations showed that the evolution of the actual climate system fell midway between the FAR and the SAR 'best estimate' projections and were within or near the upper range of projections from the TAR. Not all theories or early results are verified by later analysis. In the mid-1970s, several substances about possible global cooling appeared in the popular press, primarily motivated by analyses indicating that Northern Hemisphere temperatures had decreased during the previous three decades.

In the peer-reviewed literature by Bryson and Dittberner reported that increases in carbon dioxide should be associated with a decrease in global temperatures. When challenged by Woronko, Bryson and Dittberner explained that the cooling projected by their model was due to aerosols produced by the same combustion that caused the increase in CO_2.

However, because aerosols remain in the atmosphere only a short time compared to CO_2, the results were not applicable for long-term climate change projections.

This example of a prediction of global cooling is a classic illustration of the selfcorrecting nature of Earth science. The scientists involved were reputable researchers who followed the accepted paradigm of publishing in scientific journals, submitting their methods and results to the scrutiny of their peers and responding to legitimate criticism.

Fig. The Complexity of Climate Models has Increased over the Last Few Decades.

A recurring theme throughout this stage is that climate science in recent decades has been characterised by the increasing rate of advancement of research in the field and by the notable evolution of scientific methodology and tools, including the models and observations that support and enable the research.

During the last four decades, the rate at which scientists have added to the body of knowledge of atmospheric and oceanic processes has accelerated dramatically. As scientists incrementally increase the totality of knowledge, they publish their results in peer-reviewed journals. Between 1965 and 1995, the number of substances published per year in atmospheric science journals tripled.

Focusing more narrowly, Stanhill found that the climate change science literature grew approximately exponentially with a doubling time of 11 years for the period 1951 to 1997.

Furthermore, 95 per cent of all the climate change science literature since 1834 was published after 1951. Because science is cumulative, this represents considerable growth in the knowledge of climate processes and in the complexity of climate research. An important example of this is the additional physics incorporated in climate models over the last several decades, as showed in Figure.

The additional physics incorporated in the models are shown pictorially by the different features of the modelled world. As a result of the cumulative nature of science, climate science today is an interdisciplinary synthesis of countless tested and proven physical processes and principles painstakingly compiled and verified over several centuries of detailed laboratory measurements, observational experiments and theoretical analyses; and is now far more wide-ranging and physically comprehensive than was the case only a few decades ago.

ASSESSMENTS OF CLIMATE CHANGE AND UNCERTAINTIES

The WMO and the United Nations Environment Programme established the IPCC in 1988 with the assigned role of assessing the scientific, technical and socioeconomic information relevant for understanding the risk of humaninduced climate change.

The original 1988 mandate for the IPCC was extensive:

- Identification of uncertainties and gaps in our present knowledge with regard to climate changes and its potential impacts, and preparation of a plan of action over the short-term in filling these gaps;
- Identification of information needed to evaluate policy implications of climate change and response strategies;
- Review of current and planned national/international policies related to the greenhouse gas issue;

- Scientific and environmental assessments of all aspects of the greenhouse gas issue and the transfer of these assessments and other relevant information to governments and intergovernmental organisations to be taken into account in their policies on social and economic development and environmental programmes.'

The IPCC is open to all members of UNEP and WMO. It does not directly support new research or monitor climate-related data. However, the IPCC process of synthesis and assessment has often inspired scientific research leading to new findings. The IPCC has three Working Groups and a Task Force. Working Group I assesses the scientific aspects of the climate system and climate change, while Working Groups II and III assess the vulnerability and adaptation of socioeconomic and natural systems to climate change, and the mitigation options for limiting greenhouse gas emissions, respectively. The Task Force is responsible for the IPCC National Greenhouse Gas Inventories Programme.

This brief history focuses on WGI and how it has described uncertainty in the quantities presented. A main activity of the IPCC is to provide on a regular basis an assessment of the state of knowledge on climate change, and this volume is the fourth such Assessment Report of WGI. The IPCC also prepares Special Reports and Technical Documents on topics for which independent scientific information and advice is deemed necessary, and it supports the United Nations Framework Convention on Climate Change through its work on methodologies for National Greenhouse Gas Inventories. The FAR played an important role in the discussions of the Intergovernmental Negotiating Committee for the UNFCCC. The UNFCCC was adopted in 1992 and entered into force in 1994.

It provides the overall policy framework and legal basis for addressing the climate change issue. The WGI FAR was completed under the leadership of Bert Bolin and John Houghton in a plenary at Windsor, UK in May 1990. It made a persuasive, but not quantitative, case for anthropogenic interference with the climate system. Most conclusions from the FAR were non-quantitative and remain valid today. For example, in terms of the greenhouse gases, 'emissions resulting from human activities are substantially increasing the atmospheric concentrations of the greenhouse gases: CO2, CH4, CFCs, N2O'. On the other hand, the FAR did not foresee the phase-out of CFCs, missed the importance of biomass-burning aerosols and dust to climate and stated that unequivocal detection of the enhanced greenhouse effect was more than a decade away.

The latter two areas highlight the advance of climate science and in particular the merging of models and observations in the new field of detection and attribution. The Policymakers Summary of the WGI FAR gave a broad overview of climate change science and its Executive Summary separated key findings into areas of varying levels of confidence ranging from 'certainty' to providing an expert 'judgement'. Much of the summary is not quantitative. Similarly, scientific uncertainty is hardly mentioned; when ranges are given, as in the projected temperature increases of 0.2°C to 0.5°C per decade, no probability or likelihood is assigned to explain the range. In discussion of the climate sensitivity to doubled atmospheric CO_2 concentration, the combined subjective and objective criteria are explained: the range of model results was 1.9°C to 5.2°C; most were close to 4.0°C; but the newer model results were lower; and hence the best estimate was 2.5°C with a range of 1.5°C to 4.5°C.

The likelihood of the value being within this range was not defined. However, the importance of identifying those areas where climate scientists had high confidence was recognised in the Policymakers Summary. The Supplementary Report re-evaluated the RF values of the FAR and included the new IPCC scenarios for future emissions, designated IS92a–f. It also included updated stages on climate observations and modelling. The treatment of scientific uncertainty remained as in the FAR. For example, the calculated increase in global mean surface temperature since the 19th century was given as 0.45°C ± 0.15°C, with no quantitative likelihood for this range. The SAR, under Bert Bolin and John Houghton and Gylvan Meira Filho was planned with and coupled to a preliminary Special Report that contained intensive stages on the carbon cycle, atmospheric chemistry, aerosols and radiative forcing.

The WGI SAR culminated in the government plenary in Madrid in November 1995. The most cited finding from that plenary, on attribution of climate change, has been consistently reaffirmed by subsequent research: 'The balance of evidence suggests a discernible human influence on global climate'. The SAR provided key input to the negotiations that led to the adoption in 1997 of the Kyoto Protocol to the UNFCCC. Uncertainty in the WGI SAR was defined in a number of ways. The carbon cycle budgets used symmetric plus/minus ranges explicitly defined as 90 per cent confidence intervals, whereas the RF bar chart reported a 'mid-range' bar along with a plus/minus range that was estimated largely on the spread of published values.

The likelihood, or confidence interval, of the spread of published results was not given. These uncertainties were additionally modified by a declaration that the confidence of the RF being within the specified range was indicated by a stated confidence level that ranged from 'high' to 'very low'. Due to the difficulty in approving such a long draft in plenary, the Summary for Policy Makers became a short document with no figures and few numbers. The use of scientific uncertainty in the SPM was thus limited and similar to the FAR: a range in the mean surface temperature increase since 1900 was given as 0.3°C to 0.6°C with no explanation as to likelihood of this range. While the underlying report showed projected future warming for a range of different climate models, the Technical Summary focused on a central estimate.

The IPCC Special Report on Aviation and the Global Atmosphere was a major interim assessment involving both WGI and WGIII and the Scientific Assessment Panel to the Montreal Protocol on Substances that Deplete the Ozone Layer. It assessed the impacts of civil aviation in terms of climate change and global air quality as well as looking at the effect of technology options for the future fleet. It was the first complete assessment of an industrial sub-sector. The summary related aviation's role relative to all human influence on the climate system: 'The best estimate of the radiative forcing in 1992 by aircraft is 0.05 W m–2 or about 3.5 per cent of the total radiative forcing by all anthropogenic activities.' The authors took a uniform approach to assigning and propagating uncertainty in these RF values based on mixed objective and subjective criteria. In addition to a best value, a two-thirds likelihood interval is given. This interval is similar to a one-sigma normal error distribution, but it was explicitly noted that the probability distribution outside this interval was not evaluated and might not have a normal distribution. A bar chart with 'whiskers' showing the components and total RF for aviation in 1992 appeared in the SPM.

The TAR, under Robert Watson and John Houghton and Ding YiHui was approved at the government plenary in Shanghai in January 2001. The predominant summary statements from the TAR WGI strengthened the SAR's attribution statement: 'An increasing body of observations gives a collective picture of a warming world and other changes in the climate system', and 'There is new and stronger evidence that most of the warming observed over the last 50 years is attributable to human activities.'

The TAR Synthesis Report combined the assessment reports from the three Working Groups. By combining data on global and regional climate change, the Synthesis Report was able to strengthen the conclusion regarding human influence: 'The Earth's climate system has demonstrably changed on both global and regional scales since the pre-industrial era, with some of these changes attributable to human activities'.

In an effort to promote consistency, a guidance document on uncertainty was distributed to all Working Group authors during the drafting of the TAR. The WGI TAR made some effort at consistency, noting in the SPM that when ranges were given they generally denoted 95 per cent confidence intervals, although the carbon budget uncertainties were specified as ±1 standard deviation. The range of 1.5°C to 4.5°C for climate sensitivity to atmospheric CO_2 doubling was reiterated but with no confidence assigned; however, it was clear that the level of scientific understanding had increased since that same range was first given in the Charney et al. report. The RF bar chart noted that the RF components could not be summed and that the 'whiskers' on the RF bars each meant something different. Another failure in dealing with uncertainty was the projection of 21st-century warming: it was reported as a range covering

- Six Special Report on Emission Scenarios emissions scenarios and
- Nine atmosphere-ocean climate models using two grey envelopes without estimates of likelihood levels.

The full range of 1.4°C to 5.8°C is a much-cited finding of the WGI TAR but the lack of discussion of associated likelihood in the report makes the interpretation and useful application of this result difficult.

THE EARTH'S GREENHOUSE EFFECT

The realisation that Earth's climate might be sensitive to the atmospheric concentrations of gases that create a greenhouse effect is more than a century old. Fleming and Weart provided an overview of the emerging science. In terms of the energy balance of the climate system, Edme Mariotte noted in 1681 that although the Sun's light and heat easily pass through glass and other transparent materials, heat from other sources does not. The ability to generate an artificial warming of the Earth's surface was demonstrated in simple greenhouse experiments such as Horace Benedict de Saussure's experiments in the 1760s using a 'heliothermometre' to provide an early analogy to the greenhouse effect.

It was a conceptual leap to recognise that the air itself could also trap thermal radiation. In 1824, Joseph Fourier, citing Saussure, argued 'the temperature [of the Earth] can be augmented by the interposition of the atmosphere, because heat in the state of light finds less resistance in penetrating the air, than in repassing into the air when converted into non-luminous heat'. In 1836, Pouillit followed up on Fourier's ideas and argued 'the atmospheric stratum...exercises a greater absorption upon the terrestrial than on the solar rays'. There was still no understanding of exactly what substance in the atmosphere was responsible for this absorption. In 1859, John Tyndall identified through laboratory experiments the absorption of thermal radiation by complex molecules. He noted that changes in the amount of any of the radiatively active constituents of the atmosphere such as water or CO_2 could have produced 'all the mutations of climate which the researches of geologists reveal'. In 1895, Svante Arrhenius followed with a climate prediction based on greenhouse gases, suggesting that a 40 per cent increase or decrease in the atmospheric abundance of the trace gas CO_2 might trigger the glacial advances and retreats. One hundred years later, it would be found that CO_2 did indeed vary by this amount between glacial and interglacial periods. However, it now appears that the initial climatic change preceded the change in CO_2 but was enhanced by it. G. S. Callendar solved a set of equations linking greenhouse gases and climate change. He found that a doubling of atmospheric CO_2 concentration resulted in an increase in the mean global temperature of 2°C, with considerably more warming at the poles, and linked increasing fossil fuel combustion with a rise in CO_2 and its greenhouse effects: 'As man is now changing the composition of the atmosphere at a rate which must be very exceptional on the geological time scale, it is natural to seek for the probable effects of such a change. From the best laboratory observations it appears that the principal result of increasing atmospheric carbon dioxide... would be a gradual increase in the mean temperature of the colder regions of the Earth.' In 1947, Ahlmann reported a 1.3°C warming in the North Atlantic sector of the Arctic since the 19th century and mistakenly believed this climate variation could be explained entirely by greenhouse gas warming.

Similar model predictions were echoed by Plass in 1956: 'If at the end of this century, measurements show that the carbon dioxide content of the atmosphere has risen appreciably and at the same time the temperature has continued to rise throughout the world, it will be firmly established that carbon dioxide is an important factor in causing climatic change'. In trying to understand the carbon cycle, and specifically how fossil fuel emissions would change atmospheric CO_2, the interdisciplinary field of carbon cycle science began.

One of the first problems addressed was the atmosphere-ocean exchange of CO_2. Revelle and Suess explained why part of the emitted CO_2 was observed to accumulate in the atmosphere rather than being completely absorbed by the oceans. While CO_2 can be mixed rapidly into the upper layers of the ocean, the time to mix with the deep ocean is many centuries. By the time of the TAR, the interaction of climate change with the oceanic circulation and biogeochemistry was projected to reduce the fraction of anthropogenic CO_2 emissions taken up by the oceans in the future, leaving a greater fraction in the atmosphere.

In the 1950s, the greenhouse gases of concern remained CO_2 and H_2O, the same two identified by Tyndall a century earlier. It was not until the 1970s that other greenhouse gases–CH_4, N_2O and CFCs–were widely recognised. By the 1970s, the importance of aerosol-cloud effects in reflecting sunlight was known and atmospheric aerosols were being proposed as climate-forcing constituents. Charlson and others built a consensus that sulphate aerosols were, by themselves, cooling the Earth's surface by directly reflecting sunlight. Moreover, the increases in sulphate aerosols were anthropogenic and linked with the main source of CO_2, burning of fossil fuels. Thus, the current picture of the atmospheric constituents driving climate change contains a much more diverse mix of greenhouse agents.

PAST CLIMATE OBSERVATIONS, ASTRONOMICAL THEORY AND ABRUPT CLIMATE CHANGES

Throughout the 19th and 20th centuries, a wide range of geomorphology and palaeontology studies has provided new insight into the Earth's past climates, covering periods of hundreds of millions of years. The Palaeozoic Era, beginning 600 Ma, displayed evidence of both warmer and colder climatic conditions than the present; the Tertiary Period was generally warmer; and the Quaternary Period showed oscillations between glacial and interglacial conditions. Louis Agassiz developed the hypothesis that Europe had experienced past glacial ages, and there has since been a growing awareness that long-term climate observations can advance the understanding of the physical mechanisms affecting climate change. The scientific study of one such mechanism–modifications in the geographical and temporal patterns of solar energy reaching the Earth's surface due to changes in the Earth's orbital parametres–has a long history. The pioneering contributions of Milankovitch to this astronomical theory of climate change are widely known, and the historical review of Imbrie and Imbrie calls attention to much earlier contributions, such as those of James Croll, originating in 1864.

The pace of palaeoclimatic research has accelerated over recent decades. Quantitative and well-dated records of climate fluctuations over the last 100 kyr have brought a more comprehensive view of how climate changes occur, as well as the means to test elements of the astronomical theory. By the 1950s, studies of deep-sea cores suggested that the ocean temperatures may have been different during glacial times. Ewing and Donn proposed that changes in ocean circulation actually could initiate an ice age. In the 1960s, the works of Emiliani and Shackleton showed the potential of isotopic measurements in deepsea sediments to help explain Quaternary changes. In the 1970s, it became possible to analyse a deep-sea core time series of more than 700 kyr, thereby using the last reversal of the Earth's magnetic field to establish a dated chronology. This deep-sea observational record clearly showed the same periodicities found in the astronomical forcing, immediately providing strong support to Milankovitch's theory.

Ice cores provide key information about past climates, including surface temperatures and atmospheric chemical composition. The bubbles sealed in the ice are the only available samples of these past atmospheres. The first deep ice cores from Vostok in Antarctica provided additional evidence of the role of astronomical forcing. They also revealed a highly correlated evolution of temperature changes and atmospheric composition, which was subsequently confirmed over the past 400 kyr and now extends to almost 1 Myr. This discovery drove research to understand the causal links between greenhouse gases and climate change.

The same data that confirmed the astronomical theory also revealed its limits: a linear response of the climate system to astronomical forcing could not explain entirely the observed fluctuations of rapid ice-age terminations preceded by longer cycles of glaciations. The importance of other sources of climate variability was heightened by the discovery of abrupt climate changes. In this context, 'abrupt' designates regional events of large amplitude, typically a few degrees celsius, which occurred within several decades–much shorter than the thousand-year time scales that characterise changes in astronomical forcing. Abrupt temperature changes were first revealed by the analysis of deep ice cores from Greenland.

Oeschger et al. recognised that the abrupt changes during the termination of the last ice age correlated with cooling in Gerzensee and suggested that regime shifts in the Atlantic Ocean circulation were causing these widespread changes. The synthesis of palaeoclimatic observations by Broecker and Denton invigourated the community over the next decade. By the end of the 1990s, it became clear that the abrupt climate changes during the last ice age, particularly in the North Atlantic regions as found in the Greenland ice cores, were numerous, indeed abrupt and of large amplitude.

They are now referred to as Dansgaard-Oeschger events. A similar variability is seen in the North Atlantic Ocean, with north-south oscillations of the polar front and associated changes in ocean temperature and salinity. With no obvious external forcing, these changes are thought to be manifestations of the internal variability of the climate system. The importance of internal variability and processes was reinforced in the early 1990s with analysis of records with high temporal resolution. New ice cores new ocean cores from regions with high sedimentation rates, as well as lacustrine sediments and cave stalagmites produced additional evidence for unforced climate changes, and revealed a large number of abrupt changes in many regions throughout the last glacial cycle.

Long sediment cores from the deep ocean were used to reconstruct the thermohaline circulation connecting deep and surface waters and to demonstrate the participation of the ocean in these abrupt climate changes during glacial periods. By the end of the 1990s, palaeoclimate proxies for a range of climate observations had expanded greatly. The analysis of deep corals provided indicators for nutrient content and mass exchange from the surface to deep water, showing abrupt variations characterised by synchronous changes in surface and deep-water properties.

Precise measurements of the CH_4 abundances in polar ice cores showed that they changed in concert with the Dansgaard-Oeschger events and thus allowed for synchronisation of the dating across ice cores. The characteristics of the antarctic temperature variations and their relation to the Dansgaard-Oeschger events in Greenland were consistent with the simple concept of a bipolar seesaw caused by changes in the thermohaline circulation of the Atlantic Ocean.

This work underlined the role of the ocean in transmitting the signals of abrupt climate change. Abrupt changes are often regional, for example, severe droughts lasting for many years have changed civilizations, and have occurred during the last 10 kyr of stable warm climate. This result has altered the notion of a stable climate during warm epochs, as previously suggested by the polar ice cores. The emerging picture of an unstable oceanatmosphere system has opened the debate of whether human interference through greenhouse gases and aerosols could trigger such events. Palaeoclimate reconstructions cited in the FAR were based on various data, including pollen records, insect and animal remains, oxygen isotopes and other geological data from lake varves, loess, ocean sediments, ice cores and glacier termini.

These records provided estimates of climate variability on time scales up to millions of years. A climate proxy is a local quantitative record that is interpreted as a climate variable using a transfer function that is based on physical principles and recently observed correlations between the two records. The combination of instrumental and proxy data began in the 1960s with the investigation of the influence of climate on the proxy data, including tree rings, corals and ice cores. Phenological and historical data are also a valuable source of climatic reconstruction for the period before instrumental records became available. Such documentary data also need calibration against instrumental data to extend and reconstruct the instrumental record. With the development of multi-proxy reconstructions, the climate data were extended not only from local to global, but also from instrumental data to patterns of climate variability. Most of these reconstructions were at single sites and only loose efforts had been made to consolidate records.

Mann et al. made a notable advance in the use of proxy data by ensuring that the dating of different records lined up. Thus, the true spatial patterns of temperature variability and change could be derived, and estimates of NH average surface temperatures were obtained. The Working Group I WGI FAR noted that past climates could provide analogues. Fifteen years of research since that assessment has identified a range of variations and instabilities in the climate system that occurred during the last 2 Myr of glacial-interglacial cycles and in the super-warm period of 50 Ma. These past climates do not appear to be analogues of the immediate future, yet they do reveal a wide range of climate processes that need to be understood when projecting 21stcentury climate change.

SOLAR VARIABILITY AND THE TOTAL SOLAR IRRADIANCE

Measurement of the absolute value of total solar irradiance is difficult from the Earth's surface because of the need to correct for the influence of the atmosphere. Langley attempted to minimise the atmospheric effects by taking measurements from high on Mt. Whitney in California, and to estimate the correction for atmospheric effects by taking measurements at several times of day, for example, with the solar radiation having passed through different atmospheric pathlengths. Between 1902 and 1957, Charles Abbot and a number of other scientists around the globe made thousands of measurements of TSI from mountain sites. Values ranged from 1,322 to 1,465 W m-2, which encompasses the current estimate of 1,365 W m-2. Foukal et al. deduced from Abbot's daily observations that higher values of TSI were associated with more solar faculae.

In 1978, the Nimbus-7 satellite was launched with a cavity radiometre and provided evidence of variations in TSI. Additional observations were made with an active cavity radiometre on the Solar Maximum Mission, launched in 1980. Both of these missions showed that the passage of sunspots and faculae across the Sun's disk influenced TSI. At the maximum of the 11-year solar activity cycle, the TSI is larger by about 0.1 per cent than at the minimum. The observation that TSI is highest when sunspots are at their maximum is the opposite of Langley's hypothesis. As early as 1910, Abbot believed that he had detected a downward trend in TSI that coincided with a general cooling of climate. The solar cycle variation in irradiance corresponds to an 11-year cycle in radiative forcing which varies by about 0.2 W m-2. There is increasingly reliable evidence of its influence on atmospheric temperatures and circulations, particularly in the higher atmosphere. Calculations with three-dimensional models suggest that the changes in solar radiation could cause surface temperature changes of the order of a few tenths of a degree celsius. For the time before satellite measurements became available, the solar radiation variations can be inferred from cosmogenic isotopes and from the sunspot number.

Naked-eye observations of sunspots date back to ancient times, but it was only after the invention of the telescope in 1607 that it became possible to routinely monitor the number, size and position of these 'stains' on the surface of the Sun. Throughout the 17th and 18th centuries, numerous observers noted the variable concentrations and ephemeral nature of sunspots, but very few sightings were reported between 1672 and 1699. This period of low solar activity, now known as the Maunder Minimum, occurred during the climate period now commonly referred to as the Little Ice Age. There is no exact agreement as to which dates mark the beginning and end of the Little Ice Age, but from about 1350 to about 1850 is one reasonable estimate. During the latter part of the 18th century, Wilhelm Herschel noted the presence not only of sunspots but of bright patches, now referred to as faculae, and of granulations on the solar surface.

He believed that when these indicators of activity were more numerous, solar emissions of light and heat were greater and could affect the weather on Earth. Heinrich Schwabe published his discovery of a '10-year cycle' in sunspot numbers. Samuel Langley compared the brightness of sunspots with that of the surrounding photosphere. He concluded that they would block the emission of radiation and estimated that at sunspot cycle maximum the Sun would be about 0.1 per cent less bright than at the minimum of the cycle, and that the Earth would be 0.1°C to 0.3°C cooler. These satellite data have been used in combination with the historically recorded sunspot number, records of cosmogenic isotopes, and the characteristics of other Sun-like stars to estimate the solar radiation over the last 1,000 years. These data sets indicated quasi-periodic changes in solar radiation of 0.24 to 0.30 per cent on the centennial time scale.

These values have recently been re-assessed. The TAR states that the changes in solar irradiance are not the major cause of the temperature changes in the second half of the 20th century unless those changes can induce unknown large feedbacks in the climate system. The effects of galactic cosmic rays on the atmosphere and those due to shifts in the solar spectrum towards the ultraviolet range, at times of high solar activity, are largely unknown. The latter may produce changes in tropospheric circulation via changes in static stability resulting from the interaction of the increased UV radiation with stratospheric ozone. More research to investigate the effects of solar behaviour on climate is needed before the magnitude of solar effects on climate can be stated with certainty.

CLIMATE SYSTEM

The key to understanding global climate change is to first understand what global climate is, and how it operates. At the planetary scale, the global climate is regulated by how much energy the Earth receives from the Sun. However, the global climate is also affected by other flows of energy which take place within the climate system itself.

This global climate system is made up of the atmosphere, the oceans, the ice sheets (cryosphere), living organisms (biosphere) and the soils, sediments and rocks (geosphere), which all affect, to a greater or less extent, the movement of heat around the Earth's surface. The atmosphere plays a crucial role in the regulation of Earth's climate. The atmosphere is a mixture of different gases and aerosols (suspended liquid and solid particles) collectively known as air.

Air consists mostly of nitrogen (78 per cent) and oxygen (21 per cent). However, despite their relative scarcity, the so-called greenhouse gases, including carbon dioxide and methane, have a dramatic effect on the amount of energy that is stored within the atmosphere, and consequently the Earth's climate. These greenhouse gases trap heat within the lower atmosphere that is trying to escape to space, and in doing so, make the surface of the Earth hotter. This heat trapping is called the natural greenhouse effect, and keeps the Earth 33°C warmer than it would otherwise be. In the last 200 years, man-made emissions of greenhouse gases have enhanced the natural greenhouse effect, which may be causing global warming.

The atmosphere however, does not operate as an isolated system. Flows of energy take place between the atmosphere and the other parts of the climate system, most significantly the world's oceans. For example, ocean currents move heat from warm equatorial latitudes to colder polar latitudes. Heat is also transferred via moisture. Water evaporating from the surface of the oceans stores heat which is subsequently released when the vapour condenses to form clouds and rain. The significance of the oceans is that they store a much greater quantity of heat than the atmosphere. The top 200 metres of the world's oceans store 30 times as much heat as the atmosphere. Therefore, flows of energy between the oceans and the atmosphere can have dramatic effects on the global climate.

The world's ice sheets, glaciers and sea ice, collectively known as the cryosphere, have a significant impact on the Earth's climate. The cryosphere includes Antarctica, the Arctic Ocean, Greenland, Northern Canada, Northern Siberia and most of the high mountain ranges throughout the world, where sub-zero temperatures persist throughout the year. Snow and ice, being white, reflect a lot of sunlight, instead of absorbing it. Without the cryosphere, more energy would be absorbed at the Earth's surface rather than reflected, and consequently the temperature of the atmosphere would be much higher. All land plants make food from the photosynthesis of carbon dioxide and water in the presence of sunlight. Through this utilisation of carbon dioxide in the atmosphere, plants have the ability to regulate the global climate. In the oceans, microscopic plankton utilise carbon dioxide dissolved in seawater for photosynthesis and the manufacture of their tiny carbonate shells.

The oceans replace the utilised carbon dioxide by "sucking" down the gas from the atmosphere. When the plankton die, their carbonate shells sink to the seafloor, effectively locking away the carbon dioxide from the atmosphere. Such a "biological pump" reduces by at least four-fold the atmospheric concentration of carbon dioxide, significantly weakening the Earth's natural greenhouse effect, and reducing the Earth's surface temperature.

CLIMATE MODELLING

Climate models attempt to simulate the behaviour of the climate system. The ultimate objective is to understand the key physical, chemical and biological processes which govern climate. Through understanding the climate system, it is possible to: obtain a clearer picture of past climates by comparison with empirical observation, and; predict future climate change. Models can be used to simulate climate on a variety of spatial and temporal scales. Sometimes one may wish to study regional climates; at other times global-scale climate models, which simulate the climate of the entire planet, will be desired.

Henderson-Sellers and Robinson, Henderson-Sellers and McGuffie and Schneider provide detailed introductory discussions of the methods and techniques involved in climate modelling.

There are three major sets of processes which must be considered when constructing a climate model:

- *Radiative*: The transfer of radiation through the climate system (e.g. absorption, reflection);
- *Dynamic*: The horizontal and vertical transfer of energy (e.g. advection, convection, diffusion);
- *Surface process*: Inclusion of processes involving land/ocean/ice, and the effects of albedo, emissivity and surface-atmosphere energy exchanges.

These are the processes fundamental to the behaviour of the climate system that were discussed.

The basic laws and other relationships necessary to model the climate system are expressed as a series of equations. These equations may be empirical in derivation based on relationships observed in the real world, they may be primitive equations which represent theoretical relationships between variables, or they may be a combination of the two. Solving the equations is usually achieved by finite difference methods; it is therefore important to consider the model resolution, in both time and space i.e. the time step of the model and the horizontal/vertical scales.

Climate is the long-term statistical expression of short-term weather. Climate can be defined qualitatively as "expected weather" or quantitatively by statistical expressions such as central tendencies and variances. The overall distribution of climatological parametres, bounded by weather extremes, defines the climatic variability. Changes in climate can be defined by the differences between average conditions at two separate times. Climate may vary in different ways and over different time scales. Variations may be periodic quasi-periodic or non-periodic.

This guide represents an up-to-date review of climate change. Throughout, the focus has essentially been on global climate change, although reference to regional scale climatic change has been made if and when necessary. On their own each is a broadly self-contained discussion of a specific sub-issue of importance. The purpose of the guide is to serve as a first resort for undergraduates, applicable to first, second and third years in environmental sciences, geography and related disciplines. It is sufficient in depth and rigour for final year students but equally valuable as an information resource for those in their "pre-finals" years. It is not a complete guide but serves to review and illustrate the key factors of climate change over time and space. It contains an extensive list of references which the student is urged to consult.

The authors recognise that the study of global climate change is a rapidly developing discipline, and recent empirical evidence of climate change, and modelling evidence of the causes of climate change may have been omitted. In this sense, the guide represents a snapshot in time of the current understanding of climate change, which is open to future development. The opportunity therefore exists for readers of this guide to comment upon its content, both regarding improvements to the existing material and new ideas for additional entries. Comments and responses regarding the Global Climate Change Student Guide should be sent to The Global Climate Change Information Programme at the address on the front cover.

SIMPLIFYING THE CLIMATE SYSTEM

All models must simplify what is a very complex climate system. This is in part due to the limited understanding that exists of the climate system, and partly the result of computational restraints. Simplification may be achieved in terms of spatial dimensionality, space and time resolution, or through parametreisation of the processes that are simulated.

The simplest models are zero order in spatial dimension. The state of the climate system is defined by a single global average. Other models include an ever increasing dimensional complexity, from 1-D, 2-D and finally to 3-D models. Whatever the spatial dimension of a model, further simplification takes place in terms of spatial resolution. There will be a limited number of, for example, latitude bands in a 1-D model, and a limited number of gridpoints in a 2-D model. The time resolution of climate models varies substantially, from minutes to years depending on the nature of the models and the problem under investigation.

In order to preserve computational stability, spatial and temporal resolution have to be linked. This can pose serious problems when systems with different equilibrium time scales have to interact as a very different resolution in space and time may be needed. Parametreisation involves the inclusion of a process as a simplified (sometimes semi-empirical) function rather than an explicit calculation from first principles. Subgridscale phenomena such as thunderstorms, for example, have to be parametreised as it is not possible to deal with these explicitly. Other processes may be parametreised to reduce the amount of computation required.

Certain processes may be omitted from the model if their contribution is negligible on the time scale of interest. For example, there is no need to consider the role of deep ocean circulation whilst modelling changes over time scales of years to decades. Some models may handle radiative transfers in great detail but neglect or parametreise horizontal energy transport. Other models may provide a 3-D representation but contain much less detailed radiative transfer information.

Given their stage of development, and the limitations imposed by incomplete understanding of the climate system and computational constraints, climate models cannot yet be considered as predictive tools of future climate change. They can, however, offer a valuable window on the workings of the climate system, and of the processes that have influenced both past and present climate.

MODELLING THE CLIMATIC RESPONSE

The ultimate purpose of a model is to identify the likely response of the climate system to a change in any of the parametres and processes which control the state of the system. The climate response occurs in order to restore equilibrium within the climate system. For example, the climate system may be perturbed by the radiative forcing associated with an increase in carbon dioxide (a greenhouse gas) in the atmosphere. The aim of the model is then to assess how the climate system will respond to this perturbation, in an attempt to restore equilibrium.

The nature of the model can be one of two modes. In equilibrium mode, no account is taken of the energy storage processes which control the evolution of the climate response with time. It is assumed that the climate response occurs instantaneously following the system perturbation. The inclusion of energy storage processes allows the model to be run in transient mode, simulating the development of the climate response with time. Generally, both equilibrium and transient models will be run twice, once in a control run with no forcing, then over the test run including forcing and perturbation of the climate system. (For a further discussion of the concepts of equilibrium and transient climate responses, the reader may wish to refer back to.) The climate sensitivity and the role of feedback are critical parametres whatever the model formulation.

In the most complex models, the climate sensitivity will be calculated explicitly through simulations of the processes involved. In simpler models this factor is parametreised by reference to the range of values suggested by the more complex models. This approach, whereby more sophisticated models are nested in less complex models, is common in the field of climate modelling.

EVIDENCE FOR CLIMATE CHANGE

During the most recent antiquity, scientists have been able to construct evidence of climate change from collected information on temperature, rainfall and other weather variables from measuring stations all over the world. The oldest time series of climate data is a temperature record from central England beginning in the 17th century.

Most instrumental records however, date back only to the 19th century. Despite this relatively short period of data collection, by careful mathematical analysis scientists have been able to demonstrate that the Earth's surface has warmed on average by about 0.6°C during the 20th century. Such warming, it is believed, is related largely to mankind's pollution of the atmosphere by greenhouse gases.

Of course, it is now known that the Earth's climate has changed many times before mankind began altering the global environment, and before we began collecting information about the weather. Knowledge of palaeoclimate changes has come from the reconstruction of indirect or proxy sources of information. For climate changes that have occurred during the last 2,000 years or so, historical records may be used if they contain structure of particular weather phenomena that occurred at the time the records were written. Other forms of proxy data come from the study of natural phenomena which are climate-dependent.

The growth of tree rings, for example, is dependent on climate, and tree ring sizes and wood densities may be used to reconstruct climates at the time of growth. The oldest surviving trees are about 4,000 years old. Older wood from dead trees has been used to reconstruct past climates back about 10,000 years. To reconstruct the climate of the last Ice Age, scientists have had to turn to the chemical and physical properties of ice, collected from ice cores drilled in Greenland and Antarctica. As snow is laid down and compacted it stores a record of the current climate. The longest cores recovered are several kilometres long. Such cores maintain a record of climate changes back over 100,000 years to the pervious interglacial warm period. Palaeoclimatologists have also been able to use sea sediments to reconstruct climate during the last Ice Age and many before it, stretching back millions of years.

To investigate the oldest climate changes, some stretching back many hundreds of millions of years, geologists have to use clues from rocks, once laid down as sediment at the bottom of oceans, compressed and then uplifted to be exposed on the continents. Such geological evidence has been used to confirm ideas that continental drift and mountain building can change the Earth's climate over hundreds of millions of years.

FEEDBACK

The Earth's climate responds to changes in the amount of energy stored by the climate system, and in particular when the global energy balance between incoming energy from the Sun and outgoing heat from the Earth is upset. When"climate forcing" upsets the balance of the Earth's climate in this manner, it responds not only to the process forcing the climate to change, but also to feedback effects which can either augment or diminish the original influence.

Today, the Earth is warming up. Global warming, it is believed, is largely the result of man-made pollution of the atmosphere with excess greenhouse gases. Globally, the average surface temperature has increased by about 0.6°C during the last 100 years. In the polar regions however, regional warming has been considerably greater. In some parts of Antarctica and northern Russia, temperatures have increased by about 2°C in only 50 years. Many scientists think that such rapid warming in these ice-covered parts of the world is a consequence of the ice-albedo feedback effect.

Ice, being white, reflects a lot of sunlight and keeps the surface colder for longer. Highly reflective surfaces like ice have high albedos. As an enhanced greenhouse effect warms the surface of the Earth, some of the ice at high latitudes melts, exposing either bare ground or ocean, both of which have lower albedos (or reflectivities) than ice. With a lower albedo, the exposed surfaces reflect less sunlight, with more sunlight being absorbed. This causes a further rise in surface temperature, and in turn a further melting of ice. Here, the climatic response to primary greenhouse heating acts as a secondary climate forcing that augments the initial climate change.

Negative feedback occurs when the response to primary changes acts in the opposite direction to that of the initial climate forcing. Negative feedback reduces the climate response to initial causes of climate change. The formation of clouds in a greenhouse-heated world may cause a negative feedback. A warmer atmosphere will contain more moisture, and consequently more clouds. Clouds reflect a lot of sunlight and may help to reduce the amount of global warming due to man-made greenhouse gas emissions.

Over the longer term, changes in the shape of the Earth's orbit around the Sun are believed to influence natural global climate variations that are evident in the palaeoclimatic records over tens and hundreds of thousand of years.

It is clear however, that climate feedback effects have augmented the differences in global climate between the Ice Ages and the warmer interglacial periods, including changes in ocean circulation and atmospheric composition.

GLOBAL ENERGY BALANCE

The global energy balance is the balance between incoming energy from the Sun and outgoing heat from the Earth. The global energy balance regulates the state of the Earth's climate, and modifications to it as a result of natural and man-made climate forcing, cause the global climate to change.

Energy released from the Sun as electromagnetic radiation has a temperature of approximately 6000°C. At this temperature, electromagnetic radiation is emitted as shortwave light and ultraviolet energy. Electromagnetic radiation travels across space at the speed of light. When it reaches the Earth's, some is reflected back to space by clouds, some is absorbed by the atmosphere, and some is absorbed at the Earth's surface.

The Earth releases a lot of energy it has received from the Sun back to space. However, since the Earth is much cooler than the Sun, its radiating energy is longer wavelength infrared energy or heat. Sometimes, we can indirectly see heat radiation, for example as heat shimmers rising from a tarmac road on a hot sunny day. The energy received by the Earth from the Sun balances the energy lost by the Earth back into space. In this way, the Earth maintains a stable average temperature and therefore a stable climate (although of course differences in climate exist at different locations around the world).

The Earth atmosphere contains a number of greenhouse gases, which affect the Sun-Earth energy balance. The average global temperature is in fact 33°C higher than it should be. Greenhouse gases absorb electromagnetic radiation at some wavelengths but allow radiation at other wavelengths to pass through unimpeded. The atmosphere is mostly transparent in the visible light, but significant blocking (through absorption) of ultraviolet radiation by the ozone layer, and infrared radiation by greenhouse gases, occurs.

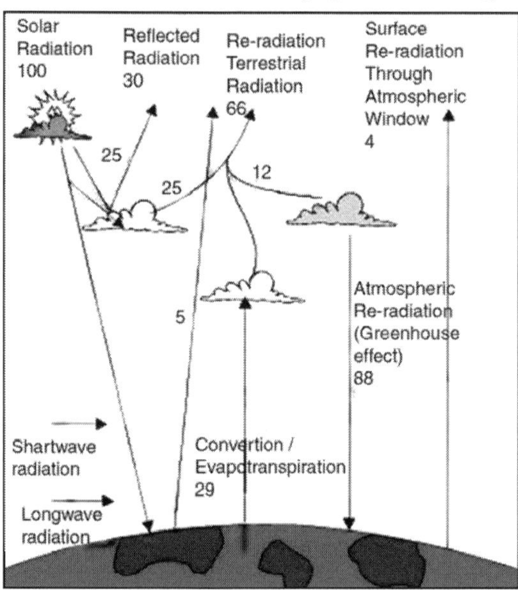

Fig. The Amount of Energy the Earth Receives from the Sun is Balanced by the Amout it loses back to Space.

The absorption of infrared radiation trying to escape from the Earth back to space is particularly important to the global energy balance. Such energy absorption by the greenhouse gases heats the atmosphere, and so the Earth stores more energy near its surface than it would if there was no atmosphere. The average surface temperature of the moon, about the same distance as the Earth from the Sun, is -18°C. The moon, of course, has no atmosphere. By contrast, the average surface temperature of the Earth is 15°C. This heating effect is called the natural greenhouse effect.

CLIMATE CONSTRUCTION FROM INSTRUMENTAL DATA

Contemporary climate change may be studied by constructing records of values (daily, monthly and annual) which have been obtained with standard equipment. The instruments must be properly installed in suitable places, carefully maintained and conscientiously observed. The instruments should be exposed in such a manner to ensure that a representative or homogeneous measurement of the climate in question is made. The concept of homogeneity is discussed further in. It is not possible to measure the climate *per se*, but only the individual elements of climate. A climate element is any one of the various properties or conditions of the atmosphere which together specify the physical state of the climate at a given place, for a particular period of time. The most commonly measured element is temperature.

MEASUREMENT OF CLIMATE ELEMENTS

Measurement of Temperature

Many surface air temperature records extend back to the middle part of the last century. The measurement of the surface air temperature is essentially the same now as it was then, using a mercury-in-glass thermometre, which can be calibrated accurately and used down to -39°C, the freezing point of mercury. For lower temperatures, mercury is usually substituted by alcohol. Maximum and minimum temperatures measured during specified time periods, usually 24 hours, provide useful information for the construction and analysis of temperature time series. Analysis involves the calculation of averages and variances of the data and the identification, using various statistical techniques, of periodic variations, persistence and trends in the time series. Observations of temperature from surface oceans are also collected in order to construct time series. In recent decades, much effort has also been directed towards the measurement of temperature at different levels in the atmosphere. There are now two methods of measuring temperatures at different altitudes: the conventional radiosonde network; and the microwave-sounding unit (MSU) on the TIROS-N series of satellites. The conventional network extends back to 1958 and the MSU data to 1979.

Temperature is a valuable climate element in climate observation because it directly provides a measure of the energy of the system under inspection. For example, a global average temperature† reveals information about the energy content of the Earth-atmosphere system. A higher temperature would indicate a larger energy content. Changes in temperature indicate changes in the energy balance, the causes of which were discussed. Variations in temperature are also subject to less variability than other elements such as rainfall and wind. In addition, statistical analysis of temperature time series is often less complex than that associated with other series. Perhaps most importantly of all, our own perception of the state of the climate are intimately linked to temperature.

MEASUREMENT OF RAINFALL

Rainfall is measured most simply by noting periodically how much has been collected in an exposed vessel since the time of the last observation. Care must be taken to avoid underestimating rainfall due to evaporation of the collected water and the effects of wind. Time series can be constructed and analysis performed in a similar manner to those of temperature.

The measurement of global rainfall offers an indirect or qualitative assessment of the energy of the Earth-atmosphere system. Increased heat storage will increase the rate of evaporation from the oceans (due to higher surface temperatures). In turn, the enhanced levels of water vapour in the atmosphere will intensify global precipitation. Rainfall is, however, subject to significant temporal and spatial variability, and the occurrence of extremes, and consequently, analysis of time series is more complex.

MEASUREMENT OF HUMIDITY

The amount of water vapour in the air can be described in at least 5 ways, in terms of:

- The water-vapour pressure;
- The relative humidity;
- The absolute humidity
- The mixing ratio
- The dewpoint.

A full account of these definitions may be found in Linacre. The standard instrument for measuring humidity is a psychrometre. This is a pair of identical vertical thermometres, one of which has the bulb kept wet by means of a muslin moistened by a wick dipped in water. Evaporation from the wetted bulb lowers its temperature below the air temperature (measured by the dry bulb thermometre). The difference between the two measured values is used to calculate the air's water-vapour pressure, from which the other indices of humidity can be determined.

MEASUREMENT OF WIND

Wind is usually measured by a cup anemometre which rotates about a vertical axis perpendicular to the direction of the wind. The exposure of wind instruments is important any obstruction close by will affect measurements. Wind direction is also measured by means of a vane, accurately balanced about a truly vertical axis, so that it does not settle in any particular direction during calm conditions.

HOMOGENEITY

Non-climatic influences - inhomogeneities - can and do affect climatic observations. Any analyst using instrumental climate data must first assess the quality of the observations. A numerical series representing the variations of a climatological element is called homogeneous if the variations are caused only by fluctuations in weather and climate.

Leaving aside the misrecording of data, the most important causes of inhomogeneity are:

- Changes in instrument, exposure and measuring technique (for example, when more technologically advanced equipment is introduced);
- Changes in station location (i.e. when equipment is moved to a new site);
- Changes in observation times and methods used to calculate daily averages; and
- Changes in the station environment, particularly urbanisation (for example, the growth of a city around a pre-existing meteorological station).

When assessing the homogeneity of a climate record, there are three major sources of information: the variations evident in the record itself; the station history; and nearby station data. Visual examination and statistical analysis of the station record may reveal evidence of systematic changes or unusual behaviour which suggest inhomogeneity. For example, there may be a step-change in the mean, indicating a change in station location. A steady trend may indicate a progressive change in the station environment, such as urbanisation. An extreme value may be due to a typing error.

Often these inhomogeneities may be difficult to detect and other evidence is needed to confirm their presence. One source of evidence is the station history, referred to as metadata. The station history should include details of any changes of location of the station, changes in instrumentation or changes in the timing and nature of observation. Very often, though, actual correction factors to observation data containing known inhomogeneities will be difficult to calculate, and in these cases, the record may have to be rejected.

The third approach to homogenisation involves empirical comparisons between stations close to each other. Over time scales of interest in climate change studies, nearby stations (i.e. within 10km of each other) should be subject to similar changes in monthly, seasonal and annual climate. The only differences should be random. Any sign of systematic behaviour in the differences (e.g. a trend or step-change) would suggest the presence of inhomogeneities. In light of the foregoing discussion on homogeneity, careful attention has to be paid to eliminating sources of non-climatic error when constructing large-scale record, such as the global surface air temperature time series. This, and similar records including sea surface temperatures, rely on the collection of millions of individual observations from a huge network made up of thousands of climate stations.

A number of these have been reviewed by Jones*et al.*. The effects of urbanisation (the artificial warming associated with the growth of towns and cities around monitoring sites) were considered to be the greatest source of inhomogeneity, but even this, it was concluded, accounts for at most a 0.05°C warming (or 10 per cent of the observed warming) over the last 100 years. Conrad and Pollak (1962), Folland *et al.* (1990), and Jones *et al.* provide useful references investigating the problems of homogeneity and data reliability of instrumental records of climate data.

CLIMATE CHANGE AND ECOSYSTEMS

Unchecked global warming could affect most terrestrial ecoregions. Increasing global temperature means that ecosystems will change; some species are being forced out of their habitats because of changing conditions, while others are flourishing. Secondary effects of global warming, such as lessened snow cover, rising sea levels, and weather changes, may influence not only human activities but also the ecosystem.

For the IPCC Fourth Assessment Report, experts assessed the literature on the impacts of climate change on ecosystems. Rosenzweig et al. concluded that over the last three decades, human-induced warming had likely had a discernable influence on many physical and biological systems. Schneider et al. concluded, with very high confidence, that regional temperature trends had already affected species and ecosystems around the world. With high confidence, they concluded that climate change would result in the extinction of many species and a reduction in the diversity of ecosystems.

- *Terrestrial ecosystems and biodiversity*: With a warming of 3°C, relative to 1990 levels, it is likely that global terrestrial vegetation would become a net source of carbon. With high confidence, Schneider et al. concluded that a global mean temperature increase of around 4°C by 2100 would lead to major extinctions around the globe.
- *Marine ecosystems and biodiversity*: With very high confidence, Schneider et al. concluded that a warming of 2°C above 1990 levels would result in mass mortality of coral reefs globally.
- *Freshwater ecosystems*: Above about a 4°C increase in global mean temperature by 2100, Schneider et al. concluded, with high

Confidence, that many freshwater species would become extinct.

Studying the association between Earth climate and extinctions over the past 520 million years, scientists from the University of York write, "The global temperatures predicted for the coming centuries may trigger a new 'mass extinction event', where over 50 per cent of animal and plant species would be wiped out." Many of the species at risk are Arctic and Antarctic fauna such as polar bears and Emperor Penguins. In the Arctic, the waters of Hudson Bay are ice-free for three weeks longer than they were thirty years ago, affecting polar bears, which prefer to hunt on sea ice. Species that rely on cold weather conditions such as gyrfalcons, and Snowy Owls that prey on lemmings that use the cold winter to their advantage may be hit hard. Marine invertebrates enjoy peak growth at the temperatures they have adapted to, regardless of how cold these may be, and cold-blooded animals found at greater latitudes and altitudes generally grow faster to compensate for the short growing season.

Warmer-than-ideal conditions result in higher metabolism and consequent reductions in body size despite increased foraging, which in turn elevates the risk of predation. Indeed, even a slight increase in temperature during development impairs growth efficiency and survival rate in rainbow trout. Rising temperatures are beginning to have a noticeable impact on birds, and butterflies have shifted their ranges northward by 200 km in Europe and North America. Plants lag behind, and larger animals' migration is slowed down by cities and roads. In Britain, spring butterflies are appearing an average of 6 days earlier than two decades ago.

A 2002 article in Nature surveyed the scientific literature to find recent changes in range or seasonal behaviour by plant and animal species. Of species showing recent change, 4 out of 5 shifted their ranges towards the poles or higher altitudes, creating "refugee species". Frogs were breeding, flowers blossoming and birds migrating an average 2.3 days earlier each decade; butterflies, birds and plants moving towards the poles by 6.1 km per decade. A 2005 study concludes human activity is the cause of the temperature rise and resultant changing species behaviour, and links these effects with the predictions of climate models to provide validation for them. Scientists have observed that Antarctic hair grass is colonizing areas of Antarctica where previously their survival range was limited.

Mechanistic studies have documented extinctions due to recent climate change: McLaughlin et al. documented two populations of Bay checkerspot butterfly being threatened by precipitation change. Parmesan states, "Few studies have been conducted at a scale that encompasses an entire species" and McLaughlin et al. agreed "few mechanistic studies have linked extinctions to recent climate change." Daniel Botkin and other authors in one study believe that projected rates of extinction are overestimated. Many species of freshwater and saltwater plants and animals are dependent on glacier-fed waters to ensure a cold water habitat that they have adapted to. Some species of freshwater fish need cold water to survive and to reproduce, and this is especially true with Salmon and Cutthroat trout. Reduced glacier run-off can lead to insufficient stream flow to allow these species to thrive. Ocean krill, a cornerstone species, prefer cold water and are the primary food source for aquatic mammals such as the Blue Whale. Alterations to the ocean currents, due to increased freshwater inputs from glacier melt, and the potential alterations to thermohaline circulation of the worlds oceans, may affect existing fisheries upon which humans depend as well.

The white lemuroid possum, only found in the mountain forests of northern Queensland, has been named as the first mammal species to be driven extinct by global warming. The White Possum has not been seen in over three years. These possums cannot survive extended temperatures over 30 °C which occurred in 2005. A final expedition to uncover any surviving White Possums is scheduled for 2009.

FORESTS

Pine forests in British Columbia have been devastated by a pine beetle infestation, which has expanded unhindered since 1998 at least in part due to the lack of severe winters since that time; a few days of extreme cold kill most mountain pine beetles and have kept outbreaks in the past naturally contained. The infestation, which has killed about half of the province's lodgepole pines is an order of magnitude larger than any previously recorded outbreak and passed via unusually strong winds in 2007 over the continental divide to Alberta. An epidemic also started, be it at a lower rate, in 1999 in Colourado, Wyoming, and Montana. The United States forest service predicts that between 2011 and 2013 virtually all 5 million acres of Colourado's lodgepole pine trees over five inches in diametre will be lost.

As the northern forests are a carbon sink, while dead forests are a major carbon source, the loss of such large areas of forest has a positive feedback on global warming. In the worst years, the carbon emission due to beetle infestation of forests in British Columbia alone approaches that of an average year of forest fires in all of Canada or five years worth of emissions from that country's transportation sources.

Besides the immediate ecological and economic impact, the huge dead forests provide a fire risk. Even many healthy forests appear to face an increased risk of forest fires because of warming climates. The 10-year average of boreal forest burned in North America, after several decades of around 10,000 km^2 has increased steadily since 1970 to more than 28,000 km^2 annually. Though this change may be due in part to changes in forest management practices, in the western U.S., since 1986, longer, warmer summers have resulted in a fourfold increase of major wildfires and a sixfold increase in the area of forest burned, compared to the period from 1970 to 1986. A similar increase in wildfire activity has been reported in Canada from 1920 to 1999.

Forest fires in Indonesia have dramatically increased since 1997 as well. These fires are often actively started to clear forest for agriculture. They can set fire to the large peat bogs in the region and the COreleased by these peat bog fires has been estimated, in an average year, to be 15 per cent of the quantity of COproduced by fossil fuel combustion.

MOUNTAINS

Mountains cover approximately 25 per cent of earth's surface and provide a home to more than one-tenth of global human population. Changes in global climate pose a number of potential risks to mountain habitats. Researchers expect that over time, climate change will affect mountain and lowland ecosystems, the frequency and intensity of forest fires, the diversity of wildlife, and the distribution of water.

Studies suggest that a warmer climate in the United States would cause lower-elevation habitats to expand into the higher alpine zone. Such a shift would encroach on the rare alpine meadows and other high-altitude habitats. High-elevation plants and animals have limited space available for new habitat as they move higher on the mountains in order to adapt to long-term changes in regional climate.

Changes in climate will also affect the depth of the mountains snowpacks and glaciers. Any changes in their seasonal melting can have powerful impacts on areas that rely on freshwater run-off from mountains. Rising temperature may cause snow to melt earlier and faster in the spring and shift the timing and distribution of run-off. These changes could affect the availability of freshwater for natural systems and human uses.

ECOLOGICAL PRODUCTIVITY

- A document by Smith and Hitz, it is reasonable to assume that the relationship between increased global mean temperature and ecosystem productivity is parabolic. Higher carbon dioxide concentrations will favourably affect plant growth and demand for water. Higher temperatures could initially be favourable for plant growth. Eventually, increased growth would peak then decline.
- The IPCC, a global average temperature increase exceeding 1.5–2.5°C would likely have a predominantly negative impact on ecosystem goods and services, *e.g.*, water and food supply.
- Research done by the Swiss Canopy Crane Project suggests that slow-growing trees only are stimulated in growth for a short period under higher CO_2 levels, while faster growing plants like liana benefit in the long term. In general, but especially in rainforests,

this means that liana become the prevalent species; and because they decompose much faster than trees their carbon content is more quickly returned to the atmosphere. Slow growing trees incorporate atmospheric carbon for decades.

CLIMATE CHANGE AND MITIGATION

Climate change mitigation is action to decrease the intensity of radiative forcing in order to reduce the potential effects of global warming. Mitigation is distinguished from adaptation to global warming, which involves acting to tolerate the effects of global warming. Most often, climate change mitigation scenarios involve reductions in the concentrations of greenhouse gases, either by reducing their sources or by increasing their sinks.

Scientific consensus on global warming, together with the precautionary principle and the fear of abrupt climate change is leading to increased effort to develop new technologies and sciences and carefully manage others in an attempt to mitigate global warming. Most means of mitigation appear effective only for preventing further warming, not at reversing existing warming. The Stern Review identifies several ways of mitigating climate change. These include reducing demand for emissions-intensive goods and services, increasing efficiency gains, increasing use and development of low-carbon technologies, and reducing fossil fuel emissions.

The energy policy of the European Union has set a target of limiting the global temperature rise to 2 °C compared to preindustrial levels, of which 0.8 °C has already taken place and another 0.5–0.7 °C is already committed. The 2 °C rise is typically associated in climate models with a carbon dioxide-equivalent concentration of 400–500 ppm by volume; the current level of carbon dioxide alone is 383 ppm by volume, and rising at 2 ppm annually. Hence, to avoid a very likely breach of the 2 °C target, CO_2 levels would have to be stabilised very soon; this is generally regarded as unlikely, based on current programmes in place to date. The importance of change is showed by the fact that world economic energy efficiency is presently improving at only half the rate of world economic growth.

CLIMATE MODEL

Global climate models use math - alot of math - to describe how the atmosphere, the oceans, the land, living things, ice, and energy from the Sun affect each other and Earth's climate. Thousands of climate researchers use global climate models to better understand how global changes such as increasing greenhouses gases or decreasing Arctic sea ice will affect the Earth. The models are used to look hundreds of years into the future, so that we can predict how our planet's climate will likely change.

There are various types of climate models. Some focus on certain things that affect climate such as the atmosphere or the oceans. Models that look at few variables of the climate system may be simple enough to run on a personal computer. Other models take into account many factors of the atmosphere, biosphere, geosphere, hydrosphere, and cryosphere to model the entire Earth system. They take into account the interactions and feedbacks between these different parts of the planet. Earth is a complex place and so many of these models are very complex too. They include so many math calculations that they must be run on supercomputers, which can do the calculations quickly. All climate models must make some assumptions about how the Earth works, but in general, the more complex a model, the more factors it takes into account, and the fewer assumptions it makes. At the National Centre for Atmospheric Research, researchers work with complex models of the Earth's climate system. Their Community Climate System Model is so complex that it requires about three trillion math calculations to simulate a single day on planet Earth. It can take thousands of hours for the supercomputer to run the model. The model output, typically many gigabytes large, is analysed by researchers and compared with other model results and with observations and measurement data. There are currently several other complex global climate models that are used to predict future climatic change. The most robust models are compared by the IPCC as they summarize predictions about future climate change.

THE PROGRAMME ON THE ETHICAL DIMENSIONS OF CLIMATE CHANGE

The major outcome of this meeting was the Buenos Aires Declaration on the Ethical Dimensions of Climate Change.

Objectives

The programme on the Ethical Dimensions of Climate Change seeks to:

- Facilitate express examination of ethical dimensions of climate change particularly for those issues entailed by specific positions taken by governments, businesses, NGOs, organizations, or individuals on climate change policy matters;
- Create better understanding about the ethical dimensions of climate change among policy makers and the general public;
- Assure that people around the world, including those most vulnerable to climate change, participate in any ethical enquiry about responses to climate change;
- Develop an interdisciplinary approach to enquiry about the ethical dimensions of climate change and support publications that examine the ethical dimensions of climate change;
- Make the results of scholarship on the ethical dimensions of climate change available to and accessible to policy makers, scientists, and citizen groups;

- Integrate ethical analysis into the work of other institutions engaged in climate change policy including the Intergovernmental Programme on Climate Change and the Conference of the Parties to the United Nations Conference on Climate Change.

Position

Given the severity of impact to be expected and given the likelihood that some level of important disruptions in living conditions will occur for great numbers of people due to climate change events, this group contends that there is sufficient convergence among ethical principles to make a number of concrete recommendations on how governments should act, or identify ethical problems with positions taken by certain governments, organizations, or individuals.

Facts about climate change and fundamental human rights provide the starting point for climate ethics.

WORLD CLIMATE REPORT

World Climate Report, a newsletter edited by Patrick Michaels, was produced by the Greening Earth Society, a non-profit organization created by the Western Fuels Association.

Early editions were document based; it was then transferred to a web-only format, having ceased publication as a physically based report with Volume 8 in 2002.

World Climate Report presents a scientific skeptical view of populist anthropogenic-driven mass global climate change, or as it describes, 'Global Warming Alarmism'.

However, it does not reject the concepts of global climate change or greenhouse theory in general attempting to engender itself as giving a well balanced and scientific view of the sources.

WCR says of itself:

- World Climate Report, a concise, hard-hitting and scientifically correct response to the global change reports which gain attention in the literature and popular press. As the nation's leading publication in this realm, World Climate Report is exhaustively researched, impeccably referenced, and always timely. This popular biweekly newsletter points out the weaknesses and outright fallacies in the science that is being touted as "proof" of disastrous warming. It's the perfect antidote against those who argue for proposed changes to the Rio Climate Treaty, such as the Kyoto Protocol, which are aimed at limiting carbon emissions from the United States... World Climate Report has become the definitive and unimpeachable source for what nature now calls the "mainstream skeptic" point of view..

WEST ANTARCTIC ICE SHEET

The West Antarctic Ice Sheet is the segment of the continental ice sheet that covers West Antarctica, the portion of Antarctica west of the Transantarctic Mountains. The WAIS is classified as a marine-based ice sheet, meaning that its bed lies well below sea level and its edges flow into floating ice shelves. The WAIS is bounded by the Ross Ice Shelf, the Ronne Ice Shelf, and outlet glaciers that drain into the Amundsen Sea.

It is estimated that the volume of the Antarctic ice sheet is about 25.4 million km^3, and the WAIS contains just under 10 per cent of this, or 2.2 million km^3. The weight of the ice has caused the underlying rock to sink by between 0.5 and 1 kilometres in a process known as isostatic depression. Under the force of its own weight, the ice sheet deforms and flows. The interior ice flows slowly over rough bedrock. In some circumstances, ice can flow faster in ice streams, separated by slow-flowing ice ridges. The inter-stream ridges are frozen to the bed while the bed beneath the ice streams consists of water-saturated sediments. Many of these sediments were deposited before the ice sheet occupied the region, when much of West Antarctica was covered by the ocean. The rapid ice-stream flow is a non-linear process still not fully understood; streams can start and stop for unclear reasons. When ice reaches the coast, it will continue to flow outward onto the water. The result is a large, floating shelf of ice affixed to the continent.

Large parts of the WAIS sit on a bed which is below sea level and slopes downward inland. This slope, and the low isostatic head, mean that the ice sheet is theoretically unstable: a small retreat could in theory destabilize the entire WAIS leading to rapid disintegration. Current computer models do not include the physics necessary to simulate this process, and observations do not provide guidance, so predictions as to its rate of retreat remain uncertain. This has been known for decades.

In January 2006, in a UK government-commissioned report, the head of the British Antarctic Survey, Chris Rapley, warned that this huge west Antarctic ice sheet may be starting to disintegrate. It has been hypothesised that this disintegration could raise sea levels by approximately 3.3 metres. Rapley said a previous Intergovernmental Panel on Climate Change report that played down the worries of the ice sheet's stability should be revised. "The last IPCC report characterized Antarctica as a slumbering giant in terms of climate change," he wrote. "I would say it is now an awakened giant. There is real concern." Note that the IPCC report did not use the words "slumbering giant".

Rapley said, "Parts of the Antarctic ice sheet that rest on bedrock below sea level have begun to discharge ice fast enough to make a significant contribution to sea level rise. Understanding the reason for this change is urgent in order to be able to predict how much ice may ultimately be discharged and over what timescale. Current computer models do not include the effect of liquid water on ice sheet sliding and flow, and so provide only conservative estimates of future behaviour." James E. Hansen, a senior NASA scientist who is a leading climate adviser to the US government, said the results were deeply worrying. "Once a sheet starts to disintegrate, it can reach a tipping point beyond which break-up is explosively rapid," he said.

Indications that the West Antarctic Ice Sheet is losing mass at an increasing rate come from the Amundsen Sea sector, and three glaciers in particular: the Pine Island, Thwaites and Smith Glaciers. Data reveals they are losing more ice than is being replaced by snowfall. A preliminary analysis, the difference between the mass lost and mass replaced is about 60 per cent. The melting of these three glaciers alone is contributing an estimated 0.24 millimetres per year to the rise in the worldwide sea level. There is growing evidence that this trend is accelerating: there has been a 75 per cent increase in Antarctic ice mass loss in the ten years 1996-2006, with glacier acceleration a primary cause.

Polar ice experts from the U.S. and U.K. met at the University of Texas at Austin in March, 2007 for the West Antarctic Links to Sea-Level Estimation Workshop. The experts discussed a new hypothesis that explains the observed increased melting of the West Antarctic Ice Sheet.

They proposed that changes in air circulation patterns have led to increased upwelling of warm, deep ocean water along the coast of Antarctica and that this warm water has increased melting of floating ice shelves at the edge of the ice sheet. An ocean model has shown how changes in winds can help channel the water along deep troughs on the sea floor, towards the ice shelves of outlet glaciers. The exact cause of the changes in circulation patterns is not known and they may be due to natural variability. However, this connection between the atmosphere and upwelling of deep ocean water provides a mechanism by which human induced climate changes could cause an accelerated loss of ice from WAIS. Recently published data collected from satellites support this hypothesis, suggesting that the west Antarctic ice sheet is beginning to show signs of instability.

The West Antarctic ice sheet has warmed by more than 0.1 °C/decade in the last 50 years, and is strongest in winter and spring. Although this is partly offset by fall cooling in East Antarctica, this effect is restricted to the 1980s and 1990s. The continent-wide average surface temperature trend of Antarctica is positive and significant at >0.05°C/decade since 1957.. This warming of WAIS is strongest in the Antarctic Peninsula.

MODELLING CLIMATE CHANGE

Climate models attempt to simulate the behaviour of the climate, in an attempt to understand the key physical, chemical and biological processes which govern climate. Climate models give us a better understanding of the climate system, providing us with a clearer picture of past climates by comparison with records of instrumental and palaeoclimatic observations, and enabling us to predict future climate change.

Models can be used to simulate climate on a variety of geographical scales and over different periods of time. The basic laws and other relationships necessary to model the climate are expressed as a series of mathematical equations. The climate however, is a very complex system, and supercomputers are needed for the task.

Global climate models have been used extensively to project global warming in the 21st century due to mankind's greenhouse gas pollution of the atmosphere. Estimates of future increases in greenhouse gases are inputted into the model, which then calculates how the global climate might evolve or respond in the future to the enhanced greenhouse effect. Although climate models can aid understanding in the processes which govern the climate, the confidence placed in such models should always be questioned. Critically, it should be remembered that all climate models represent a simplification of the climate system, a system that may ultimately prove to be too complex to model accurately.

Climate models must therefore be used with care and their results interpreted with due caution. Margins of uncertainty should be attached to any model projection. Results from climate models should always be validated or tested against real-world data, including both instrumental and palaeoclimatic records where available.

MOUNTAINS

There is now little doubt that the presence of mountain ranges on the Earth can dramatically influence global climate, and that episodes of mountain building through Earth Antiquity have acted as mechanisms of global climate change. The process of mountain building is very slow however, being associated with continental drift and taking tens or even hundreds of millions of years. Consequently, changes in climate as a result of such tectonic movements develop only over such long periods of time.

The temperature difference between the equator and the poles generates wind which carries heat gradually north and south from low latitudes to high latitudes on the Earth. Global atmospheric circulation patterns however, maintain a more east-west trend on account of the Earth's rotation, which deflects moving air due to the Coriolis force.

North-south orientated mountain ranges therefore, have the ability to influence such circulation patterns which regulate the global climate system. The Rocky Mountains for example, which extend along the western side of the North American continent, influence significantly the climate on their eastern flank.

Air originating from the Pacific Ocean is deflected north around them before returning southwards, bringing with it exceptionally cold temperatures in winter to much of the central US and Canada. Scientists have suggested that the uplift of the Tibetan Plateau and the Himalayas, resulting from the collision of India into Asia 20 million years ago, may have contributed to the long term global cooling that has taken place during the last 40 million years.

In addition to its effect on global atmospheric circulation, the mountain uplift exposed greater volumes of rock to the effects of chemical and physical weathering:

- During chemical weathering, carbon dioxide dissolved in rainwater helps decompose rock minerals to form bicarbonates.
- These bicarbonates are soluble and are transported via rivers downstream to the coasts where they are deposited on ocean floors as sediment.
- This locks away the carbon dioxide for millions of years, thereby reducing the amount of carbon dioxide remaining in the atmosphere. Since carbon dioxide is a greenhouse gas, increased rates of mountain building can diminish the strength of the Earth's natural greenhouse effect, contributing to global cooling.
- Air is colder at higher altitudes, where precipitation may fall as snow rather than rain. Mountain uplift increases the amount of high altitude land area and therefore the surface area covered by snow the year round. Snow, of course, has a much greater reflectivity or albedo than most other snow- and ice-free surfaces.
- The subsequent increase in the amount of reflected sunlight reduces the amount of energy absorbed at the Earth's surface, encouraging further cooling. This cooling process is known as the ice-albedo feedback effect.

OCEANS

The oceans store an immense amount of heat energy, much more than the atmosphere, and consequently play a crucial role in the regulation of the global climate. As in the atmosphere, surface ocean currents assist in the transfer of heat from low to high latitudes:

- Warm water moves poleward whilst cold water returns towards the equator.
- Energy is also transferred via moisture.
- Water evaporating from the surface of the oceans stores heat which is subsequently released when the moisture condenses to form clouds and rain.

Heat exchanges also occur vertically within the oceans, between surface water, usually the top 200 metres or so, and the deep water. Seawater in the high latitudes readily sinks, forming deep-water currents. Considerable evaporation of moisture takes place from the warm surface ocean currents as they travel towards the high latitudes. The salt which remains behind in the water after evaporation makes the water heavier or denser. As in the atmosphere the surface and deep-water currents of the world's oceans are inter-linked forming the global ocean circulation.

Scientists have proposed that changes in this global ocean circulation influence climate changes over hundreds and thousands of years. Shifts in the Earth's orbit around the Sun have been shown to be linked to changes in the Earth's climate over hundreds of thousands of years. Scientists now recognise however, that such orbital variations are on their own not enough to account for the shifts in global climate between cold Ice Ages and warm interglacials. Whilst orbital variations may indeed performance as a pacemaker to warm-cold climate transitions, additional climate feedback processes have been invoked to explain the large changes in global average temperature of up to 5°C.

Changes in the composition of the atmosphere, in particular the levels of the greenhouse gas carbon dioxide, have been shown, through reconstruction of palaeoclimatic records, to account for the colder climate of the last Ice Age. A change in ocean circulation however, has been proposed as the trigger mechanism for the transition from a warm interglacial climate to a cold Ice Age 120,000 years ago, and the switch back again 14,000 years ago. Scientists believe that at the end of the last warm interglacial period 120,000 years ago, high latitude seawater was gradually cooling as a consequence of a reduction in heat received by the Sun due to orbital variations.

Colder seawater in the high latitudes will display less deep-water circulation. Sea sediment records have revealed that this change in deep water formation is particularly evident in the Northern Hemisphere. Colder seawater loses less water through evaporation of moisture, and is therefore less salty and lighter. The lighter water finds it more difficult to sink, reducing or even shutting down completely that branch of the global ocean circulation. The warm poleward-moving surface currents too, are slowed or redirected, and consequently less heat is transferred to the polar regions.

Less heat, of course, means a colder climate, leading to the growth of ice sheets across the Northern Hemisphere landmasses and the development of a new Ice Age.

Climate models have suggested that the loss of the warm surface Gulf Stream in the North Atlantic, which keeps the climate of western Europe mild, could result in a regional drop in temperature of 6 to 8°C. In addition, such changes in ocean circulation could occur over relatively short periods, perhaps within 50 to 100 years. Rather paradoxically, scientists have now begun to speculate that global warming could threaten the present course and intensity of the Gulf Stream:

- In this climate scenario deep water formation is reduced, not by a reduction in moisture evaporation, but by a melting of ice from the Greenland ice cap.
- The influx of freshwater would reduce the saltiness of the seawater, making it lighter and more difficult to sink.
- Worryingly, such a pattern of climate change can be seen the palaeoclimatic record 11,000 years ago, at the beginning of the present interglacial warm period.
- Large volumes of melting ice from the retreating Northern Hemisphere ice sheets flowed into the North Atlantic, shut down the regional ocean circulation and cause a temporary fall in regional temperature of several degrees Celsius in just a few hundred years.

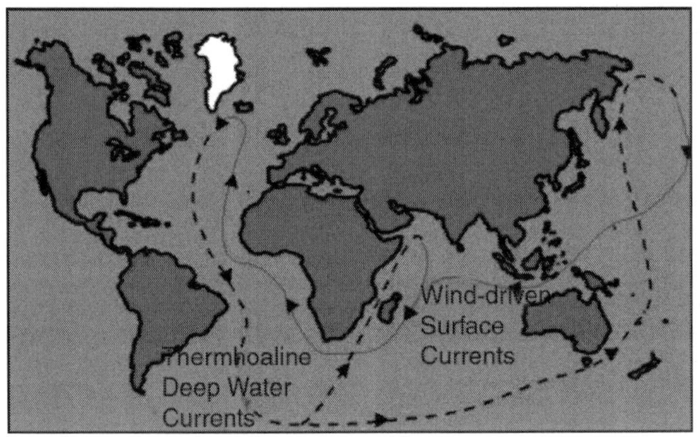

Fig. Ocean Circulation Plays and Important Role in the Regulation of the Earth's Climate.

PALAEOCLIMATE CHANGE

The period for which we have instrumental records of climate change, such as observational records of temperature and rainfall, spans only a tiny fraction of Earth Antiquity. Furthermore, although we are now concerned with global warming due to mankind's greenhouse gas pollution of the atmosphere, this contemporary climate change should be placed in the context of much longer term changes in climate that have taken place quite naturally. Prehistoric climate change is known to climatologists and Earth scientists as palaeoclimatic change.

The global climate has shifted and varied for billions of year, perhaps since the Earth first had an atmosphere. The oldest palaeoclimatic records have allowed us to reconstruct climate fairly reliably during the last 500 million years. Over this time, the global climate has moved from extensive periods of global warmth to periods of global cold several times, each lasting 100 million years or more.

Although today we are concerned about global warming, we do in fact lie in the middle of global ice-house climate, which began 40 million years ago, when the first permanent ice sheets formed on Antarctica. The change from the much warmer global climate which existed during the age of the dinosaurs, when global average temperature was perhaps 10°C higher than at present, is thought to have been caused by changes in the distribution of landmasses and the associated changes to energy redistribution throughout the climate system.

Within the long-term global icehouse climate, much shorter-term fluctuations in global climate have occurred. Relatively cold periods known as Ice Ages or glacials, each lasting roughly 100,000 years, are interspersed with much shorter warmer episodes or interglacials, lasting only 10,000 years. We now have a relatively clear record of such climatic fluctuations over the last 2 million years. Currently, the global climate lies within an interglacial. Global average temperature 20,000 years ago towards the end of the last Ice Age was some 5°C lower than today, when the north polar ice sheets were expanded to cover a considerably greater area of the continental Northern Hemisphere than is the case today. These glacial-interglacial fluctuations are believed to be driven by changes in the position of the Earth in its orbit around the Sun, and enhanced by climatic feedback processes which involve changes in ocean circulation and the greenhouse gas composition of the atmosphere.

Within the latest present interglacial period, further fluctuations in the global climate can be seen in the palaeoclimatic records and more recently in the instrumental records. During the last 1000 years, the climate has moved from a period of Medieval warmth to a"Little Ice Age" between the 16th and 19th centuries, with changes of between 0.5 and 1°C in the global average surface temperature. Although it is not clear what has caused these climatic changes, variations in the Sun's energy output, ocean circulation and the occurrence of major volcanic eruptions are believed to play a part. Most recently, we have entered a renewed period of global warming since the beginning of the 20th century that we suspect is the result of mankind's enhancement of the natural greenhouse effect through the pollution of the atmosphere. Global average temperature is now about the same as it was during the Medieval warm period, although still much lower than it was 100 million years ago during the age of the dinosaurs.

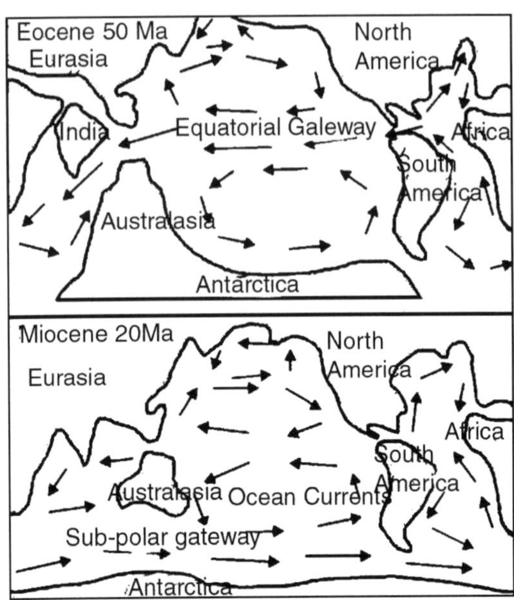

Climate change that predates the instrumental period of direct weather observations is known as palaeoclimate change. Palaeoclimatology provides a longer perspective on climate variability that can improve our understanding of the climate system, and help us to predict future climatic changes as a result of man-made global warming.

Evidence for palaeoclimatic change can be obtained by the study of natural phenomena which are climate-dependent. Such evidence comes from palaeoclimatic records. Many natural systems are dependent on climate. From these it may be possible to derive palaeoclimatic information for variables such as temperature or rainfall. Such proxy or indirect records of climate contain a climatic signal. Deciphering that signal is often a complex business.

Four types of palaeoclimatic record are commonly analysed for past climate variations. These include historical records, glaciological records, biological records and geological records. The suitability of a particular palaeoclimatic record for reconstructing a past climate will be largely dependent upon the time scale of past climate change under study. Recent climate changes during the last few thousand years can be reliably reconstructed from tree ring analyses, which often yield continuous records and provide high-resolution (annual or even seasonal) data.

Over the longer term, ice cores and sea sediments offer information about palaeoclimates stretching back hundreds of thousands of years, although the data resolution may not be as fine. Generally, the further back in time we go the greater the margins of uncertainty that will be attached to palaeoclimatic reconstructions.

PALAEOCLIMATOLOGY

Palaeoclimatology, from the Greek word"palaios", meaning"ancient", is the study of past climates and past climate change, prior the period of instrumental records. The study of palaeoclimatology can encompass much of Earth Antiquity, or at least that part of it for which reliable palaeoclimatic records are available to reconstruct palaeoclimates.

Palaeoclimatology may be distinguished from climatology and contemporary climate change, which studies present day climate and climate changes restricted to the most recent period (the last 150 years) since instrumental records of daily weather observations have become available. To reconstruct palaeoclimates, palaeoclimatologists cannot use direct observations of temperature, rainfall and other climatic variables.

Instead they use proxy records of natural phenomena which are climate-dependent. These include analyses of tree rings, ice cores, sea sediments and even rock strata exposed at the Earth's surface which may hold clues to the state of the climate millions of years ago. Climate models run on computers may also be used to test theories about possible mechanisms of palaeoclimate change.

SEA SEDIMENTS

Billions of tonnes of sediment accumulate in the ocean basins every year. The nature of such sediments may be indicative of climate conditions near the ocean surface or on the adjacent continents. Sediments are composed of both organic and inorganic materials.

The organic component of sea sediment includes the remnants of sea-dwelling microscopic plankton, which provide a record of past climate and oceanic circulation. For example, by studying the chemical composition of plankton shells, we can reveal information about past seawater temperatures, salinity (saltiness), and nutrient availability. Indeed, such techniques have been used to reconstruct ocean temperatures over the last 100 million years, and have confirmed continental drift theories of climate change that a long term global cooling has taken place since the extinction of the dinosaurs.

Most inorganic material comes from adjacent landmasses, eroded from rocks and washed down to the coast by river channels, or blown from soils, dusty plains and deserts. The nature and abundance of inorganic materials provides information about how wet or dry the nearby continents were, and the strengths and directions of winds.

SUN

For many years scientists have speculated that changes in the amount of energy given off by the Sun can influence the Earth's climate. There is no doubt that variations do occur in various characteristics of the Sun on a range of time scales. The 11-year cycle in the number of sunspots on the face of the Sun is well known. But other parameters, including the size of the Sun, vary too, and over different time scales from tens of years to thousands of years. What is less clear is whether or not these changes produce significant variations in the total energy output of the Sun. The total solar energy received by the Earth, or solar constant, has only been measured accurately since the advent of the satellite era, 40 years ago.

In addition, changes which have been detected over the past 20 years are very small in magnitude, and probably too small to account for all of the observed global warming that has occurred during this period. While changes in total solar energy may be greater on longer time scales, this is only a speculative possibility. Nevertheless, scientists have proposed that longer-term changes to the nature of the sunspot cycle may have been the cause of the Little Ice Age that occurred between the 16th and 19th centuries.

TIME

Climate changes on all time scales because there are different mechanisms operating over different periods of times which"force" the climate. Most of these mechanisms are entirely natural in origin, but over the last 20 years, concerns have grown that a man-made change in the composition of the atmosphere may be causing global warming. In general, mechanisms of climate change alter the amount of energy stored by the climate system and in particular the atmosphere. The more energy stored the warmer the climate. The longest periods of global climate change occur over hundreds of millions of years, in response to episodes of continental drift and mountain building. Within such long term cycles there exist much shorter climatic fluctuations over tens and hundreds of thousands of years, driven by changes in the Earth's orbit around the Sun, and modulated by changes in ocean circulation and atmospheric composition.

Over the shortest time scales of centuries, decades and even individual years, global climate is influenced by changes in the amount of energy from the Sun and the effects of major volcanic eruptions. The actual state of the global climate at any point in time represents an aggregate response to all the mechanisms which influence it. How the global climate responds to this"climate forcing" will depend upon the different response times of the various parts of the climate system. The overall response time of the global climate will then be determined by the interactions between the parts. As mankind is releasing more greenhouse gases into the atmosphere for example, the atmosphere is slowly trapping more heat over the years and enhancing the Earth's natural greenhouse effect.

The oceans in contact with the atmosphere however, respond much more slowly to changes in atmospheric heating, since they require a much greater quantity of energy to warm them up. A temperature increase of the oceans due to global warming may take hundreds of years. This will moderate and slow the response of the climate system to global warming, giving mankind a little breathing space with which to tackle the problem.

TREE RINGS

The study of the relationships between annual tree growth and climate is called dendroclimatology. Dendroclimatology offers a high-resolution (annual) form of palaeoclimate reconstruction for the last few thousand years. Whenever tree growth is limited directly or indirectly by some climate element, usually temperature or rainfall, and that limitation can be quantified and dated, dendroclimatology can be used to reconstruct some information about past environmental conditions. There are several subfields of dendroclimatology associated with the processing and interpretation of different tree-growth variables. Such variables include tree ring width, tree ring density, and chemical or isotopic variables.

VOLCANOES

Atmospheric pollution from major volcanic eruptions can influence the global climate over one to two years. Explosive volcanic eruptions can inject large quantities of dust and sulphur dioxide, in gaseous form, to an altitude of over 10 miles into the atmosphere (the stratosphere), where the sulphur dioxide is rapidly converted into secondary sulphuric acid aerosols. Whereas volcanic pollution from smaller eruptions, ejected only a few miles into the atmosphere, is removed within days by rain, the volcanic dust and aerosols in the stratosphere may remain for up to two years, gradually spreading over much of the globe by winds.

Volcanic pollution results in a 5 to 10 per cent reduction in direct sunlight, largely through scattering as a result of the highly reflective sulphuric acid aerosols. Large eruptions, such as the Mount Pinatubo in the Philippines which occurred in 1991, can bring about a short but noticeable global cooling of up to 0.3°C.

It is possible the longer term changes in the amount of volcanic activity on the Earth may explain incidences of longer term climate change during Earth Antiquity. 66 million years ago, at about the same time as the extinction of the dinosaurs, massive amounts of lava were being erupted on the Indian sub-continent, at that time drifting North towards Asia, having separated from Australia and Antarctica. These immense lava flows, now known as the Deccan Traps, released huge quantities of gases, including sulphur dioxide and carbon dioxide.

Theories concerning the climatic impact of such emissions vary. Some suggest that an initial severe global cooling may have occurred as result of a reduction in the amount of incoming sunlight, scattered and reflected by the secondary sulphuric acid aerosols. Others suggest that the extra volumes of carbon dioxide enhanced the Earth's greenhouse effect, causing a longer term global warming. Whatever the climatic effects were due to, it is likely that they were responsible for the mass extinction of living species that occurred at this time, including not just the death of the dinosaurs, but many other living organisms on land and in the oceans.

THE EFFECTS OF CLIMATE CHANGE ON ECOSYSTEMS

One way that scientists gain understanding of how global warming will affect ecosystems is to analyse the effects of past climate on paleoecosystems. A paleoecosystem is an ecosystem that existed in a former geologic time period. By relating vegetative cover to past climates, models can be developed. Once a reliable model is created, input variables such as CO_2 can be varied and the results analysed. An example of what the effect would be like on populations of Douglas fir in the northwestern United States if the CO_2 content in the atmosphere were double what it was before the Industrial Revolution was modeled by the U.S. Geological Survey and is shown in the illustration that follows. In a study conducted by the U.S. Global Change Research Programme in Washington, D.C., first in 2001, then updated in 2004, in trying to predict what the effects of future climate change would have on ecosystems, they concluded that"climate change has the potential to affect the structure, function, and regional distribution of ecosystems, and thereby affect the goods and services they provide." They based their conclusions on a modeling and analysis project they conducted called the Vegetation/Ecosystem Modeling and Analysis Project.

This project was used to generate future ecosystem scenarios for the conterminous United States based on model-simulated responses to both the Canadian and Hadley scenarios of climate change. The VEMAP was subsequently used in a validation exercise for a Dynamic Global Vegetation Model by Oregon State University and the U.S. Forest Service in 2008. Their MC1-DGVM was used as the input data in both the VEMAP and VINCERA. Their MC1 was run on both the VEMAP and VINCERA climate and soil input data to document how a change in the inputs can affect model outcome. The simulation results under the two sets of future climate scenarios were compared to see how different inputs can affect vegetation distribution and carbon budget projections.

The results indicated that"under all future scenarios, the interior west of the United States becomes woodier as warmer temperatures and available moisture allow trees to get established in grasslands areas. Concurrently, warmer and drier weather causes the eastern deciduous and mixed forests to shift to a more open canopy woodland or savannatype while boreal forests disappear almost entirely from the Great Lakes area by the end of the 21st century. While under VEMAP scenarios the model simulated large increases in carbon storage in a future woodier west, the drier VINCERA scenarios accounted for large carbon losses in the east and only moderate gains in the west. But under all future climate scenarios, the total area burned by wildfires increased." The similarities of the two models served to validate the VEMAP project.

The Hadley model was developed by the Hadley Centre for Climate Prediction and Research in England. Also referred to as the Met Office Hadley Centre for Climate Change, it is based at the headquarters of the Met Office in Exeter. It is the key institution in the United Kingdom for climate research. It is currently involved not only with understanding the physical, chemical, and biological processes within the climate system, but also with developing working models to explain current phenomena and to predict future climate change.

It also monitors global and national climate variability and change and strives to determine the causes of the fluctuations. The Canadian climate model was developed by the Canadian Centre for Climate Modelling and Analysis. The CCCma is a division of the Climate Research Branch of Environment Canada based out of the University of Victoria, Victoria, British Columbia. Its specific focus is on climate change and modeling. In the past nine years, the CCCma has produced three atmospheric and three atmospheric/oceanic general circulation models, making them one of the international leaders in climate change research. What they found was that over the next few decades climate change in the United States will most likely lead to increased plant productivity as a result of increasing levels of CO_2 in the atmosphere. There will also be an increase in terrestrial carbon storage for many parts of the country, especially the areas that become warmer and wetter. The southeast will most likely see reduced productivity and, therefore, a decrease in carbon storage. By the end of the 21st century, many areas of the country will have experienced changes in the distribution of vegetation. Wetter areas will see the growth of more trees; drier areas will have drier soils that will cause forested areas to die off and be replaced by savanna/grassland ecosystems.

Modeling the vegetation evolution and adaptation is more difficult. The study focused on two time periods: 2025-2034 and 2090-2099. In the near term, biogeochemical changes are expected to dominate the ecological responses. Biogeochemical responses include changes based on the natural cycles of carbon, nutrients and water. The responses are affected by changing environmental conditions such as temperature, precipitation, solar radiation, soil texture, and atmospheric CO_2. It is these natural cycles that affect carbon capture by plants with photosynthesis, soil nitrogen processes, and water transfer. These biogeochemical factors are what influence the production of vegetation. In the results from the near-term biogeochemical model, the scientists concluded that there would be an increase in CO_2. They estimate that currently the average carbon storage rate is 66/Tg/yr.

The Hadley model predicted that carbon storage rates by 2025-2034 would increase to 117/Tg/yr. The Canadian model estimated CO_2 to increase 96 Tg/yr. The Canadian model projected that the southeastern ecosystems will lose carbon in the near term because they predict the climate there will become hot and dry. The biogeography models look at the changing landscape based on changes in CO2, evapotranspiration, vegetation establishment, and competition between species, growth rates, and life cycle/mortality rates. In this model, scientists at both the Hadley and Canadian Centre for Climate Change agree that vegetation will be able to freely move from one location to another.

Changes in vegetation distribution will vary from region to region as follows:

- *Northeast*: Forests remain the dominant natural vegetation, but forest mixes will change. There will also be some increase in savannas and wetlands.
- *Southeast*: Forests remain the dominant ecosystem, but mixes change. Savannas and grasslands encroach on forests, especially towards the end of the 21st century. Drought and wildfires contribute significantly to forest destruction.
- *Midwest*: Forests remain the dominant land cover, but changes in species type occur. There will be a modest expansion of savannas and grasslands.
- *Great Plains*: Slight increase in woody vegetation.
- *West*: The areas of desert ecosystems shrink, and forest ecosystems grow.
- *Northwest*: Forested areas grow slightly.

A separate study conducted by the Canadian government predicts the following ecosystem changes as a result of changing climate over the next 100 years:

- *Coastlines*:
  - Flooding and erosion in coastal regions
  - Sea-level rise
- *Forests*:
  - Increase in pests
  - Increased levels of drought and wildfire
- *Plants and animals*:
  - Warmer temperatures could make water supplies more scarce, having a negative impact on plants and animals, not giving them time to adjust.
- *Crops*:
  - In some areas, warmer climate may allow a three- to five-week extension of the frost-free season, which could benefit commercial agriculture.

  - In other areas, drier soils and lack of water will have a negative impact on agricultural productivity.
- *Wells*:
  - The quality and quantity of drinking water may be threatened by increasing drought.
- *Harsh weather:*
  - Winter storms, floods, drought, heat waves, and tornadoes could become more frequent and severe.
- *Fisheries*:
  - Populations and ranges of species sensitive to changes in water temperature will be negatively affected.
  - Salmon harvests will be lower in the Pacific.
  - Changes in ocean currents may have a negative impact on the fisheries in the Atlantic.
- *Lakes and rivers*:
  - Water levels will decline under the influence of drought, negatively affecting drinking water quality.
  - Use of lakes for transportation, recreation, and fishing, and the ability to generate electricity may be curtailed under droughtlike conditions.
  - Other areas that may have an increase in precipitation may experience flooding, rising sea levels, and severe storms.

In February 2005, the Met Office in Exeter, England, issued a report titled"Avoiding Dangerous Climate Change." The objective of the study was to determine what levels of CO_2 were considered the tipping point for dangerous climate change with harmful effects on ecosystems and what actions could be taken now to avoid such an outcome. In the report, then prime minister Tony Blair stated:"It is now plain that the emission of greenhouse gases... is causing global warming at a rate that is unsustainable." Environment Secretary Margaret Beckett stated,"The report's conclusions would be a shock to many people.The thing that is perhaps not so familiar to members of the public... is this notion that we could come to a tipping point where change could be irreversible.

We're not talking about it happening over five minutes, of course, maybe over a thousand years, but it's the irreversibility that I think brings it home to people." The report, published by the British government, says there is only a small chance of greenhouse gas emissions being kept below"dangerous" levels. It warns that the Greenland ice sheet could melt, causing sea levels to rise by 23 feet over the next 1,000 years. It also warns that developing countries will be the hardest hit. The report also states, based on the vulnerability of many of the world's ecosystems, that the European Union (EU) has adopted a target of preventing an increase in global temperature of more than 3.3°F (2°C). Some believe even that may be too high. The report states:"Above two degrees, the risks increase very substantially," with"potentially large numbers of extinctions" and"major increases in hunger and water shortage risks... particularly in developing countries." In order to meet their goals, British scientists have advised that CO_2 levels should be stabilized at 450 parts per million (ppm) or below. Currently the atmosphere contains 380 ppm. In response to this, the British government's chief scientific adviser, Sir David King, said that was unlikely to happen. He stated,"We're going to be at 400 ppm in 10 years' time. I predict that without any delight in saying it." Myles Allen, an expert on atmospheric physics at Oxford University, said that:"Assessing a 'safe level' of carbon dioxide in the atmosphere was 'a bit like asking a doctor what's a safe number of cigarettes to smoke per day.'"

The report does conclude, however, that there are technological options available to reduce CO_2 emissions that will need to be used. The study also concluded that the biggest obstacles involved with using these new technologies, along with renewable resources of energy and"clean coal," are the current economic investments and traditionally strong bond to the oil industry, cultural attitudes that oppose change, and simple lack of awareness by many people. Various conservation organizations currently involved in the battle against global warming, such as the Union of Concerned Scientists (UCS), the Defenders of Wildlife, and the World Wildlife Fund (WWF), also support these ideas.

GLOBAL PERMAFROST CARBON CYCLE

The Permafrost Carbon Cycle is a sub-cycle of the larger global Carbon Cycle. Permafrost is defined as subsurface material that remains below 0° C for at least two consecutive years. Because permafrost soils remain frozen for long periods of time, they store large amounts of carbon and other nutrients within their frozen framework during that time. Permafrost represents a large carbon reservoir that is seldom considered when determining global terrestrial carbon reservoirs. Recent and ongoing scientific research however, is changing this view. The permafrost carbon cycle deals with the transfer of carbon from permafrost soils to terrestrial vegetation and microbes, to the atmosphere, back to vegetation, and finally back to permafrost soils through burial and sedimentation due to cryogenic processes.

Some of this carbon is transferred to the ocean and other portions of the globe through the global carbon cycle. The cycle includes the exchange of carbon dioxide and methane between terrestrial components and the atmosphere, as well as the transfer of carbon between land and water as methane, dissolved organic carbon, dissolved inorganic carbon, particulate inorganic carbon and particulate organic carbon. Soils, in general, are the largest reservoirs of carbon in terrestrial ecosystems. This is also true for soils in the Arctic that are underlain by permafrost. Determining carbon stocks in cryosols was completed using the Northern and Mid Latitudes Soil Database. Permafrost affected soils cover nearly 9 per cent of the earth's land area, yet store between 25 and 50 per cent of the soil organic carbon. These estimates show that permafrost soils are an important carbon pool. These soils not only contain large amounts of carbon, but also sequester carbon through cryoturbation and cryogenic processes. Carbon is not produced by permafrost. Organic carbon derived from terrestrial vegetation must be incorporated into the soil column and subsequently be incorporated into permafrost to be effectively stored. Because permafrost responds to climate changes slowly, carbon storage removes carbon from the atmosphere for long periods of time. Radiocarbon dating techniques reveal that carbon within permafrost is often thousands of years old.

Carbon storage in permafrost is the result of two primary processes:

- The first process that captures carbon and stores it is syngenetic permafrost growth. This process is the result of a constant active layer thickness and energy exchange between permafrost, active layer, biosphere, and atmosphere, resulting in the vertical increase of the soil surface elevation. This aggradation of soil is the result of aeolian or fluvial sedimentation and/or peat formation. Peat accumulation rates are as high as 0.5mm/yr while sedimentation may cause a rise of 0.7mm/yr. Thick silt deposits resulting from abundant loess deposition during the last glacial maximum form thick carbon-rich soils known as yedoma. As this process occurs, the organic and mineral soil that is deposited is incorporated into the permafrost as the permafrost surface rises.
- The second process responsible for storing carbon is cryoturbation, the mixing of soil due to freeze-thaw cycles. Cryoturbation moves carbon from the surface to depths within the soil profile. Frost heaving is the most common form of cryoturbation. Eventually, carbon that originates at the surface moves deep enough into the active layer to be incorporated into permafrost. When cryoturbation and the deposition of sediments act together, carbon storage rates increase.

The amount of carbon stored in permafrost soils is poorly understood. Current research activities seek to better understand the carbon content of soils throughout the soil column. Recent studies estimate that northern circumpolar permafrost soil carbon content equals approximately 1672 Pg. This estimation of the amount of carbon stored in permafrost soils is more than double the amount currently in the atmosphere. This most recent assessment of carbon content in permafrost soils breaks the soil column into three horizons, 0–30 cm, 0–100 cm, and 1–300 cm. The uppermost horizon, 0–30 cm contains approximately 191 Pg of organic carbon. The 0–100 cm horizon contains an estimated 496 Pg of organic carbon, and the 0–300 cm horizon contains an estimated 1024 Pg of organic carbon. These estimates more than doubled the previously known carbon pools in permafrost soils. Additional carbon stocks exist in yedoma, carbon rich loess deposits found throughout Siberia and isolated regions of North America, and deltaic deposits throughout the Arctic. These deposits are generally deeper than the 3 m investigated in traditional studies.

Many concerns arise because of the large amount of carbon stored in permafrost soils. Until recently, the amount of carbon present in permafrost was not taken into account in climate models and global carbon budgets. Thawing permafrost may release great quantities of old carbon stored in permafrost to the atmosphere. Carbon stored within arctic soils and permafrost is susceptible to release due to several different mechanisms. Carbon that is stored in permafrost is released back into the atmosphere as either carbon dioxide, CO_2, or methane, CH_4. Aerobic respiration releases carbon dioxide, while anaerobic respiration releases methane.

- Microbial activity releases carbon through respiration. Increased microbial decomposition due to warming conditions is believed to be major source of carbon to the atmosphere. The rate of microbial decomposition within organic soils, including thawed permafrost, depends on environmental controls. These controls include soil temperature, moisture availability, nutrient availability, and oxygen availability.
- Methane clathrate, or hydrates, occur within and below permafrost soils. Because of the low permeability of permafrost soils, methane gas is unable to migrate vertically through the soil column. As permafrost temperature increases, permeability also increases, allowing once trapped methane gas to move vertically and escape. Dissociation of gas hydrates is common along the Arctic coastline yet, estimates for dissociation of gas hydrates from terrestrial permafrost remains unclear.
- Thermokarst/permafrost degradation as a result of climate change and increased mean annual air temperatures throughout the Arctic threatens to release large quantities of carbon back into the atmosphere. The spatial extent of permafrost decreases in warming climate, releasing large amounts of stored carbon.
- As air and permafrost temperatures change, above ground vegetation also changes. Increasing temperatures facilitate the transfer of soil carbon to growing vegetation on the surface. This transfer removes carbon from the soil and relocates it to the terrestrial carbon pool where plants process, store, and respire it, moving it to the atmosphere.
- Forest fires in the boreal forests and tundra fires alter the landscape and release large quantities of stored organic carbon

into the atmosphere through combustion. As these fires burn, they remove organic matter from the surface. Removal of the protective organic mat that insulates the soil exposes the underlying soil and permafrost to increased solar radiation which in turn increases the soil temperature, active layer thickness, and changes soil moisture. Changes in the soil moisture and saturation alter the ratio of oxic to anoxic decomposition within the soil.

- Hydrologic processes remove and mobilize carbon, carrying it downstream. Mobilization occurs due to leaching, litter fall, and erosion. Mobilization is believed to be primarily due to increased primary production in the Arctic resulting in increased leaf litter entering streams and increasing the dissolved organic carbon content of the stream. Leaching of soil organic carbon from permafrost soils is also accelerated by warming climate and by erosion along river and stream banks freeing the carbon from the previously frozen soil.

Carbon is continually cycling between soils, vegetation, and the atmosphere. Currently, carbon flux from permafrost soils is minimal, however studies suggest that future warming an permafrost degradation will increase the CO_2 flux from the soils. Thaw deepens the active layer, exposing old carbon that has been in storage for decades, to centuries, to millennia. The amount of carbon that will be released from warming conditions depends on depth of thaw, carbon content within the thawed soil, and physical changes to the environment. The likelihood of the entire carbon pool mobilizing and entering the atmosphere is low despite the large volumes stored in the soil. Although temperatures are projected to rise, it does not imply complete loss of permafrost and mobilization of the entire carbon pool. Much of the ground underlain by permafrost will remain frozen even if warming temperatures increase the thaw depth or increase thermokarsting and permafrost degradation.

Warmer conditions are expected to cause spatial declines in permafrost extent and thickening of the active layer. This decline in the extent and volume of permafrost enables the mobilization of stored soil organic carbon to the biosphere and atmosphere as carbon dioxide and methane. Additionally, these changes are believed to impact ecosystems and alter the vegetation that is present on the surface. Increased carbon uptake by plants is expected to be relatively small when compared to the amount of carbon released by permafrost degradation. Tundra vegetation contains 0.4 kg of carbon per m^2 while a shift to boreal forests could increase the above ground carbon pool to 5 kg of carbon per m^2. Tundra soil however, contains ten times that amount. Additionally, a sudden and steady release of carbon dioxide and methane from permafrost soils may lead to a positive feedback cycle where warming releases carbon dioxide into the atmosphere. This carbon dioxide, a greenhouse gas, causes atmospheric concentrations to increase, causing subsequent warming. This scenario is thought to be a potential runaway climate change scenario.

BASIC COMPONENTS OF THE ECOSYSTEM

It is helpful to visualize ecosystems as consisting of four basic components: abiotic substances, producer organisms, consumer organisms and decomposer organisms.

Abiotic Components

Abiotic substances are the inorganic and organic substances not momentarily present in living organisms. These include water, carbon dioxide, oxygen, nitrogen, minerals, salts, acids, bases and the entire range of elements and compounds outside living organisms at any given point in time.

Many elements may be tightly bound in inorganic compounds, such as silicon in sandstone or aluminum in feldspar, and are unavailable to living organisms. Elements which are readily available to living organisms such as free O_2 or CO_2, or they may be in an inaccessible form such as silicon dioxide (SiO_2) in quartz, a major component of granite.

Similarly, potassium may be readily available to plants in the form of KCL in soil, but relatively unavailable in the form of $KAlSi_3O_8$ in orthoclase or monoclinic feldspar, one of the commonest of all minerals. An important property of an ecosystem which determines its productivity is the form and composition in which bioactive elements and compounds occur. For example, an ecosystem may have a substantial abundance of vital nutrients, such as nitrates and phosphates, but if they are present in relatively insoluble particulate form as they would be if linked to ferric ions, they would not be so readily available to plants as if they were in the soluble form of potassium or calcium nitrate and phosphate.

One of the most important qualities of an ecosystem is the rate of release of nutrients from solids, for this regulates the rate of function of the entire system.

Producers

Producer organisms are bacteria and plants which synthesize organic compounds. They are said to be autotrophic or self-productive, in that they take inorganic compounds and manufacture organic materials and living protoplasm from them. All green plants, including microscopic algae, are producer organisms since they exhibit photosynthesis, and some bacteria are producers since they may exhibit chemosynthesis or photosynthesis. Obviously, all life depends upon the basic productive capacity of green plants and bacteria.

Consumers

Consumer organisms are animals which utilize the organic materials directly or indirectly manufactured by plants. Consumers are unable to produce their own organic compounds for basic nutritive purposes. They are said to be heterotrophic, which means different or varied in nutritional source.

Primary consumers or herbivores directly consume the organic compounds of plants. Secondary consumers may be omnivores or carnivores which depend partially or entirely on other animals for food. Tertiary and quartenary consumers may be the second or third-stage predator, for example, a hawk feeding on a weasel which in turn consumed a mouse.

Decomposers

Decomposer organisms are bacteria and fungi which degrade organic compounds. Their nutrition is said to be saprophytic, that is, associated with rotten an ecosystem—they reduce the complex organic molecules of dead plants and animals to simpler organic compounds which can be absorbed by green plants as vital nutrients. They provide the final essential link in the cycle of life.

They are necessary for the renewal of life, for if decomposers were not active, organic compounds would become locked into complex insoluble molecules which could not be utilized as nutrients by plants.

Ecosystems involve, of course, a wide variety of life forms not specifically mentioned in the preceding paragraphs, but virtually all components of an ecosystem can be classified into producers, consumers, or decomposers. For example, parasites are merely specialized consumers.

Plant parasites feed directly on plants and are thus herbivores; animal parasites derive their nutrition from other animals, and are thus carnivores differing from predators only in the fact that they normally do not kill the host. Scavengers such as vultures are also carnivores, differing from predators by the fact that they feed on an animal after it has died from some other cause.

INCOMPLETE ECOSYSTEMS

Almost all ecosystems have all four basic components discussed above, though in some cases it is possible for incomplete ecosystems to exist. These are ecosystems lacking one or more basic components. An example of an incomplete ecosystem lacking producers is the abyssal depths of the sea where only consumers and decomposers exist.

In the realm of complete darkness green plants cannot survive. Scavengers and decomposers live on the fall-out of animals, plants and organ matter from the upper layers of the ocean. Predators might also be present to feed upon the scavengers. Hence, the ecosystem depends on extrinsic production, namely, the fall-out from upper levels. It might be possible, of course, for a few chemosynthetic bacteria to be present, but they would not produce a significant volume of organic material.

The same situation exists in caves where complete darkness prevents the growth of green plants. Again, a few chemosynthetic bacteria might be present, but they would not produce a significant amount of organic material. Practically all cave-dwelling-animals must depart from the cave, as do bats, or depend on extrinsically produced nutrients which enter the cave by flowing water or seepage.

The central core of the city might also be considered an incomplete ecosystem without producers, at least from the human standpoint. Some green plants obviously exist, but they would not supply meaning ful production for humans, sparrows, dogs, cats, etc. For all of these, the inner city requires extrinsic production and imported food. The only other alternative is for the inhabitants to leave the inner city and feed in peripheral areas. This undoubtedly occurs with starlings and pigeons, creating an ecological situation analogous to bats in a cave. The lack of production in an inner city is not due to a lack of light, but to a lack of soil and suitable substrate.

In other ways, cities may be considered incomplete ecosystems, ecologically parasitic upon the surrounding landscape. Not only do they import food, but they must also import fresh air and water. At the same time, they must export waste products—sewage, solid waste, carbon dioxide, sulphar dioxide, etc. If cities were encapsulated from their surrounding environments they would soon perish from thirst, starvation, asphyxiation, or the accumulation of waste products. In exchange for this life support, cities, of course, provide a great many economic and cultural benefits—jobs, housing, transportation, manufacturing, education, etc. So the relationship between city and landscape is vital in both directions, but it is particularly important to remember, as cities expand, that they cannot sustain themselves.

Incomplete Ecosy

stems also exist in specialized cases where producers and decomposers are present without consumers. A theoretical example would be a massive bloom of some toxic algae in an aquatic ecosystem, where the algae would create toxic conditions for zooplankton and fish and all other possible consumers. Then a process of excessive production and massive decomposition would go hand in hand. This would be a highly unstable and undesirable circumstance, but it has been known to occur, as, for example, in the red tides of Florida.

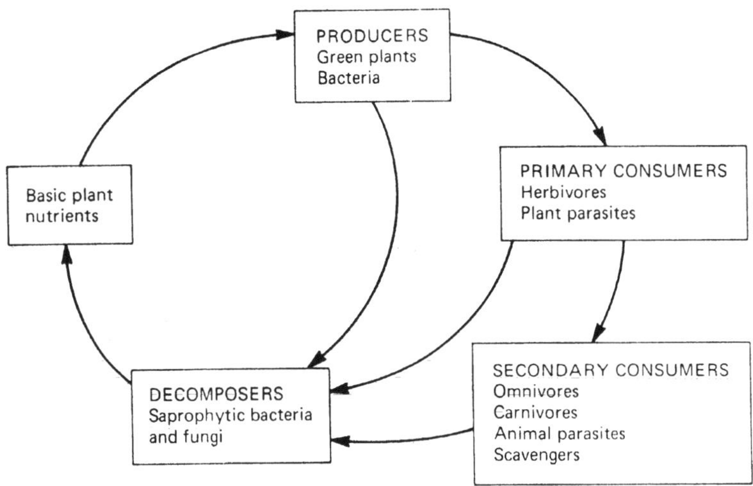

Fig. Basic Components of an Ecosystem. The Arrows Represent Some of the Major Pathways of Organic Compounds.

A third type of incomplete ecosystem might even be called an abiotic ecosystem, that is, one without living organisms, a self-contradiction in terms. They should more properly be called abiotic environments. Apollo space flights have shown so far that the moon is abiotic. Local areas on earth may be abiotic, for example, the high altitude ice plateau of Antarctica is probably devoid of living organisms over rather extensive areas. Closer home for most of us, the Copperhill basin of Tennessee is an area devoid of life, where fumes from copper smelters are toxic to all organisms within a certain downwind area. Possibly a few bacteria exist in very limited places, but for all practical purposes no plants or animals can survive.

8

Soils, Weathering, and Nutrients

SOILS

All our elements, with the exception of hydrogen and helium, came from the dying throes of large stars. Amino acids needed to build animal proteins come from autotrophic plant life.

Biological studies have shown how certain nutrients (*e.g.*, nitrogen, phosphorus, calcium, magnesium, potassium, iron, etc.) are essential for the process of protein formation. All our amino acids and nutrients eventually come to us from plant life (sometimes via the meat of plant-eating animals). Plants synthesize amino acids from the combination of *sunlight, water and soils.*

Soil is therefore of critical importance to life. Simply put: no soil, no life. We first define soil as a *dynamic natural body capable of supporting a vegetative cover.* Where there is no soil, there is no plant life and we have barren rock and/or sand. Soil is composed primarily of *weathered* materials, along with water, oxygen and organic materials. Luckily for us, soil covers most of the land surface with a fragile, thin mantle. Soil and agricultural scientists have identified a huge number of different soil types.

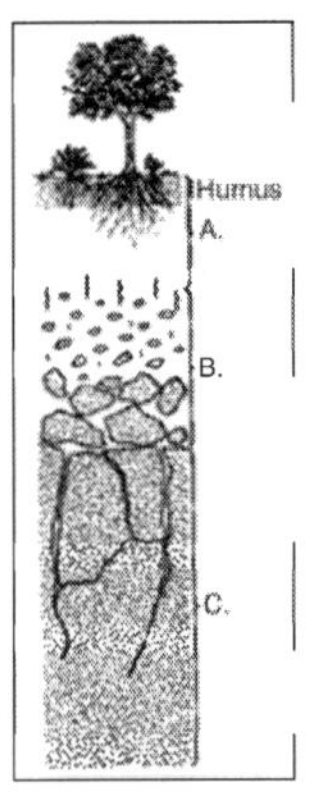

Fig. 3.1 Layers of Horizons of a Typical Soil Profile

SOIL IS LAYERED

Soil is layered into sections called "horizons". Figure 3.1 shows a typical soil profile developed on granite bedrock in a temperate region. The top horizon is composed of *humus* and contains most of the organic matter. This layer is often the darkest. The "A" horizon consists of tiny particles of decayed leaves, twigs and animal remains. The minerals in the A-horizon are mostly clays and other insoluble minerals. Minerals that dissolve in water are found at greater depths.

The "B" horizon has relatively little organic material, but contains the soluble materials that are *leached* downwards from above. The "C" horizon is slightly broken-up bedrock, typically found 1-10 meters below the surface. While this is a typical soil profile, many other types exist, depending on climate, local rock conditions and the community of organisms living nearby. The U.S. Department of Agriculture has classified 10 orders and 47 suborders of soils. If you include other subsets, there are over 60,000 types of soil. The lunar surface, which has been produced by meteoroid impacts, is not classified as a soil, but is rather given the name "regolith" (derived from the Greek words meaning cover and stone). The layered nature of soil indicates its long evolution under the effects of atmospheric and biological processes. The process that creates soil from bare rock is called "weathering". In the weathering process, the atmosphere and water interact with bare rock to slowly break it down into smaller and smaller particles. Rock climbers who encounter talus slopes (regions of pebble-life rocks that form in great conical piles at the feet of mountains) experience an intermediate step in the inexorable transition from solid granite to sand and soil. We next discuss the process whereby bare volcanic rock can be slowly turned into soils that can support life.

SOIL DEVELOPMENT

Soil forms from a complex interaction between earth materials, climate, and organisms acting over time. The brightly coloured soils of the humid tropics reflect the intense chemical reactions occurring in warm climates. The fertile prairie soils of the American Midwest evolved from the nutrient-rich organic matter left by decaying grasses. Regardless of soil characteristics, the whole process starts with the breakdown of earth material.

WEATHERING

Weathering refers to processes that physically breakdown and chemically alter earth material. *Physical weathering*, also known as *mechanical weathering*, is the breakdown of large pieces of earth material into smaller ones. Think of physical weathering as the *disintegration* of rock without changing its chemical composition. There are many ways earth material can be physically weathered. When water freezes in rock crevices it expands creating stress in the crevice. As the stress increases, the crevice widens ultimately breaking the rock. Plant roots wedge rocks apart as they grow into rock crevices too.

The shrinking and swelling by alternating heating and cooling weakens mineral bonds causing the rock to disintegrate. A very important result of physical weathering is its impact on the surface area of weathered material. When a block of earth material is broken into several smaller pieces, the amount of exposed surface increases. Examine the diagram below. A block with a width, depth, and height of 1 cm has a total surface area of 6 square centimeters.

If we break the block in half in all directions it yields eight smaller pieces all with width, height, and depth of .5 cm. Breaking the block apart creates additional exposed surfaces such that the total surface area is now 12 square centimeters. Having more total exposed surface provides more area upon which chemical reactions can take place to further weather the material. The shape of the pieces also affects the the amount of exposed surface area. Plate-like pieces have more exposed surface area than do block-like pieces.

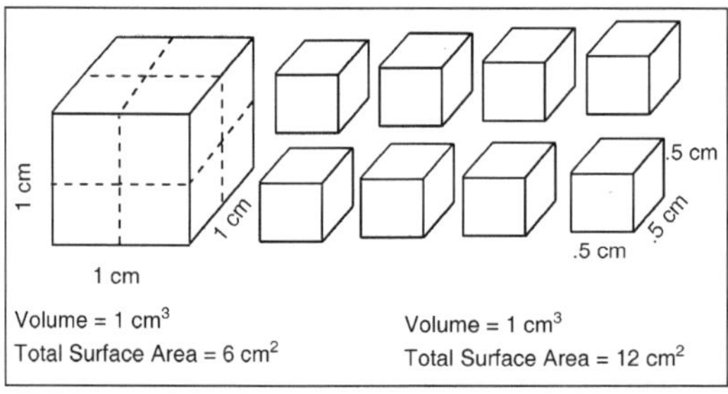

Fig. 3.3 Effect of Physical Weathering on Surface Area

Chemical weathering breaks down earth material by chemical alteration. This usually means adding a substance like water or air to the material. For instance, when oxygen is added to iron bearing minerals, *oxidation* takes places and a loose mantle of iron oxide is created (rust). *Hydrolysis* is an exchange reaction involving minerals and water. Free hydrogen (H^+) and hydroxide $(OH)^-$ ions in water replace mineral ions and drive them into solution. As a result, the mineral's structure is changed into a new form.

Hydrolysis is a common process whereby silicate minerals are weathered into a clay mineral. Think of chemical weathering as the decomposition of earth material.

The spatial variation of climate and organisms play a significant role in the weathering of earth materials. Dry locations tend to be dominated by physical weathering and moist places by chemical weathering. The type of earth material available also determines the amount of weathering that might take place.

Limestone is easily broken down where abundant rainfall and high temperatures prevail. However, limestone will remain intact in dry locations. The end result of the weathering process is the creation of a *weathered mantle*. The weathered mantle is not yet a soil until it undergoes further change. This involves the addition, transformation, translocation and removal of materials from the weathered mantle to form distinctive soil layers.

Since early geologic time, the atmosphere has interacted with the Earth's exposed crust though a process known as *weathering*. Weathering takes place through a combination of both mechanical and chemical means. We have all experienced the results of weathering first hand.

Any visit to an old cemetery find us peering at the blurred inscriptions on old marble tombstones. These inscriptions were once perfectly legible, but with the passage of time, the small fractures and cracks in the rock have made it vulnerable to attack by aqueous solutions. A dramatic example of weathering can be seen in these two images of the same 3000 year old Egyptian obelisk just before relocation to damp New York and 100 years after accelerated weathering in New York Weathering rates are obviously a strong function of climate!

Many of the original volcanic gases (*e.g.*, carbon dioxide, sulphur-bearing gases, etc.) were able to dissolve in water and produce acids. The acids, in turn, reacted with surface minerals. Later, oxygen in the atmosphere reacted with the exposed reduced materials, making the red beds discussed in an earlier lecture. Since the advent of land plants, soil and surface minerals have been exposed to relatively high concentrations of carbon dioxide maintained in soil pores as a result of decomposition and the metabolic activities of roots. The reaction of carbon dioxide with water in the soil produces carbonic acid (H_2CO_3) which determines the rate of rock weathering in most ecosystems.

$$CO_2 \text{ (gas)} + H_2O \text{ (liquid)}—H_2CO_3 \text{ (solution)}$$

Acid rain, produced by human effluents of nitrogen and sulphur-bearing gases will increase the rate of rock weathering in downwind areas. To understand why weathering occurs and why the rate of rock weathering is so dependent on climate, we need to discuss the chemistry of the process in a little more detail.

IGNEOUS ROCK WEATHERING

There is a well known expression that captures much of the story of rock weathering and illustrates the important role it has played in Earth history:

"Igneous Rocks + Acid Volatiles = Sedimentary Rocks + Salty Oceans"

What we mean by this will become clearer if we look at the details of the weathering processes. Consider a boulder or rock containing *Feldspar* minerals. Feldspar is a general term for a group of aluminosilicate minerals containing sodium, calcium, or potassium and having a lattice framework structure that makes for rigidity. Feldspars turn out to be one of the most common minerals in the Earth's crust. Feldspars are weathered through the chemical process of *hydration:*

$$K\,Al\,Si_3O_8 + H_2O \rightarrow Al_2SiO_5(OH)_4$$

In this chemical formula feldspar reacts with water to produce a kaolinite (clay). Notice that the chemical equation does not exactly balance, that is, not all the elements on the left hand side appear on the right hand side. This is because soluble elements, such as potassium (K) are *leached* out during the chemical reaction and carried away as dissolved salts. The process of leaching can perhaps best be understood by analogy with the making of coffee.

When hot water is passed over crushed coffee beans, the soluble components (making the coffee) are leached away, leaving the insoluble crushed coffee bean remnants behind.

In this way, rocks containing feldspars are weakened though the conversion of rigid feldspar to more plastic clays which do not have anything like the same structural rigidity. The process occurs at exposed surfaces of the minerals making up the rock. Through geologic time, large amounts of sedimentary r0ocks have been deposited as part of this process. In fact about 75 per cent of all exposed rocks on the Earth's surface today are of sedimentary origin and have been brought to the surface by geologic uplift.

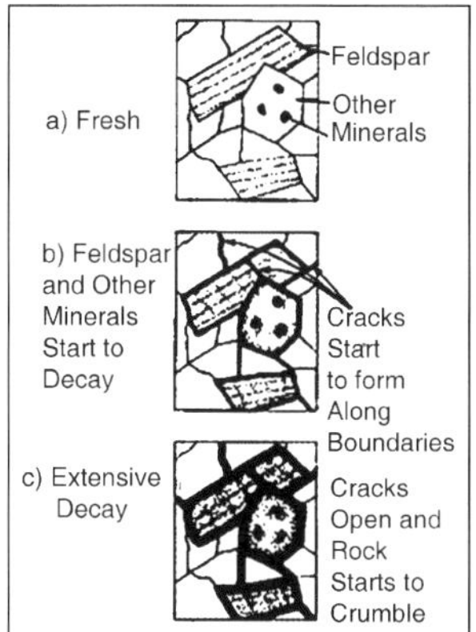

Of course, geological processes return some of the sedimentary rocks to the mantle of the Earth, where they are converted back to the primary materials under conditions of great temperature and pressure. Rock weathering is also critical for the release of biochemical elements that have no gaseous form-examples are calcium, Ca, Potassium, K, Iron, Fe, and Phosphorus, P. The latter element plays a key role in cell metabolism. Thus, we can say that weathering provides key nutrients for life through the process of leaching.

The soluble nutrients are transferred to soils layers below the immediate surface (*e.g.*, the "B" layer). This is a major reason that plants have evolved root systems-to search out these critically needed nutrients below ground.

Igneous rock weathering proceeds in stages. It is useful to picture the inside of a rock-made up of interlocking minerals of irregular shapes, each being rigid. Figure 3.5 shows a microscopic view of such an interior rock composition, with some of the grains being feldspars.

As weathering proceeds, the boundaries of the vulnerable feldspar (and other) mineral grains start to decay. As the decay proceeds, water can reach more and more feldspar surfaces and the process accelerates. This process can also be accelerated by melt-freeze cycles that force the grains apart due to the difference between the volume occupied by water and ice. We can see that weathering is due to the combined effects of chemical and mechanical decay.

The chemical and physical processes of weathering transform the igneous rock into sand and clay particles and dissolved salts. Chemical weathering can add carbon dioxide, water, and oxygen. The link provided courtesy of the National Park Service shows an example of a weathering rock. It is interesting to note that, since most of the Earth's exposed rock is of sedimentary origin and since sedimentary rocks are a by-product of weathering, most of the rocks that are weathering away in today's world are second or perhaps even third generation rocks.

That is, they originated as igneous material, became sedimentary through weathering and transport to the bottom of shallow waters, were then subject to geologic uplift to become again exposed and, finally, began to undergo weathering yet again. The natural world is full of such endless cycles.

HOW FAST DOES A ROCK DECAY

The best answer to this question is... it depends. It depends on the local environment and the type of rock. For example, an iron nail buried in the ground in Michigan will only take a year or so to decay to the point that it is easily snapped in two. Iron nails rust much more slowly in drier environments. Aluminum cans decay very slowly, even in humid climates. Glass decays even more slowly, while plastic is considered essentially non-biodegradable. It is somewhat ironic that a plastic tombstone will endure much longer than one made in marble!

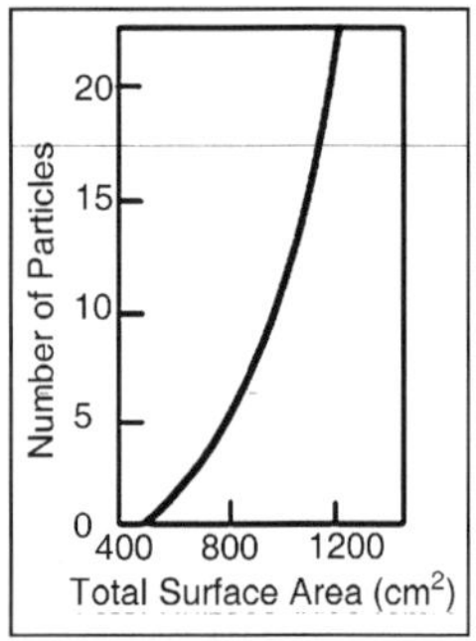

From what we have already discussed, soils themselves aid in rock decay, as do melt-freeze cycles and bacterial action. Thus, soils are a consequence of weathering, but also a factor in accelerating weathering. The production of soil is a *positive feedback* process. The following table illustrates rates of weathering for three rock types as a function of climate. As more of a rock becomes amenable to weathering, the speed of weathering increases. This can be understood if we plot the rate at which the available exposed surface area of a rock increases as the rock fragments. Figure 3.6 shows this relationship.

NUTRIENTS

Although living tissue is composed of carbon, hydrogen, and oxygen in the approximate proportion of CH_2O, as many as 23 other elements are necessary for biochemical reactions and for the growth of structural biomass.

Other important nutrients include magnesium, potassium, iron, sulphur, etc. We should note that, although carbon, nitrogen and sulphur can be obtained from the atmosphere, calcium, magnesium, potassium, iron, and phosphorus all come from rock weathering processes. The atmosphere has no store of these essential nutrients

Examples of important nutrients are:

Table. Rates of Weathering of Clean Rock Surfaces (Micro-Meters/1000Years)

Rock Type	Cold Climate	Warm, Humid Climate
Basalt	10	100
Granite	1	10
Marble	20	200

Nutrient	Role
Nitrogen	The proteins found in plants and animals contain about 16 per cent by weight of nitrogen
Phosphorus	Pis part of the important ribulose biphosphate carboxylase molecule and is part of ATP-adenosine triphosphate, the universal molecule for energy transformations
Calcium	Ca is a major structural component of the proteins forming plants and animals

As mentioned earlier, one of the main purposes of plant root systems is to get access to nutrients stored in the soil. Some plants go to enormous lengths to do this.

For example, Emiliani quotes that "A single plant of winter rye, 50 cm high, was found to have a root system consisting of 143 main roots, 35,600 secondary roots, 2.3 million tertiary roots, and 11.5 million quaternary roots! The root system was found to have a total length of 600 km and a total surface of about 250 square meters". Delivery of nutrients into plant root systems can occur by several pathways. In some cases, direct uptake in water solution occurs. Sometimes, plants actually have to protect themselves against too much nutrient intake.

Too much of a good thing can prove poisonous. An example of this can be seen in the accumulations of calcium carbonate deposits that surround the roots of some desert shrubs.

Table. Nature's Vitamin Requirements

Amount	Nutrient
100 parts	N
15 parts	P
50 parts	K
5 parts	Ca
5 parts	Mg
10 parts	S

Some nutrients, such as nitrogen, phosphorus, and potassium are often harder for roots to find and specialized (incredibly efficient) enzymes have evolved located in root membranes to seek out these scarce and needed resources. If some nutrients are not readily available, plants will grow more slowly and/or increase their root/shoot ratio.

In general, the availability of nutrients (deficit or surplus availabil ity) often controls the form of the ecosystem, determining its overall productivity, and influencing which particular set of plants come to predominate. Excess nitrogen can lead to the loss of fine root biomass and deficiencies in other nutrients.

The pool of nutrients held in the soil and vegetation is many times larger than the annual receipt of nutrients from the atmosphere and rock weathering. Thus, life *husbands* its needed nutrients on land, storing much of the total in the humus. Recycling of nutrients is critical to the productivity of natural ecosystems, although less critical to crop production, due to the availability of commercial fertilizers.

Table.Percentage of Annual Nutrient Requirement for Growth of Hardwood Forest

Process (kg/ha/yr)	N	P	K	Ca	Mg	
Total Growth Requirement	115	12	67	62	10	Atmospheric
Inputs	18	0	1	4	6	Rock Weathering
Inputs	0	13	11	34	37	Reabsorptions
(Intra-system)	31	28	4	0	2	Detritus
Turnover	69	81	86	85	87	

The data of the table came from a study of the famous Hubbard Brook ecosystem in New Hampshire. It shows how effective the soils are in storing the needed nutrients and how limited are the rates of supply from atmosphere and rock weathering.

Acid rain caused by human emissions of nitrogen and sulphur oxides leads to enhanced weathering and changes in nutrient ratios. For example, recent studies suggest that forest growth has declined in areas downwind of air pollution. Acid rain appears to increase the movement of aluminum ions, which may, in turn, reduce the intake rates of calcium and other nutrients.

In the oceans, life is also limited by the availability of nutrients. The productivity is highest on continental shelves and in regions of upwelling. Nutrients are removed from surface waters by downward sinking and are regenerated in deep waters. A paucity of nutrients limits production in the open oceans. On the other hand, marine productivity is threatened by excessive human inputs of nitrogen and phosphorus in coastal regions, where it leads to excessive algal blooms, whose subsequent decay robs deeper water of oxygen, creating “dead zones”.

SOIL EROSION

We define soil erosion as the movement of surface litter and topsoil. The forces responsible are wind and water flow. It is important to recognize at the outset that soil erosion is a natural process. It is slowed, however, by plant roots that serve to stabilize the soil. In any undisturbed ecosystem, the rate of soil loss is matched by its rate of production. In a disturbed ecosystem, however, major changes in the rate of erosion can occur. In fact, almost every activity that can be characterized as a "development" causes enhanced soil erosion. This includes farming, logging, building, grazing, off-road travel, etc. Soil erosion, discussed later in this lecture, is responsible for the loss of a great deal of our topsoil.

The problem of accelerated soil erosion is a major one. On a global scale, topsoil is eroding faster than it can be replenished in over one third of the world's croplands.

For example, in China and India combined, more than 12 million square kilometers have been severely eroded since 1945. The causes of this have been deforestation (30 per cent), overgrazing (35 per cent), and farming (28 per cent). The soil erosion problem is particularly severe if one considers that each year we must feed an extra 90 million more people, with about 25 billion tons less topsoil!

In the U.S., the situation is also of concern. The Dust Bowl of the 1930's was caused by plowing of the prairies. Before the pioneers, the soil was held in place by the long root systems of prairie grasses. In response to the dust bowl, the Soil Conservation Service (SCS) was established in 1935. The Dust Bowl was brought on by a long drought and lasted about a decade. Much was learned about soil conservation and later droughts have not caused as much damage.

The Great Plains has lost about one-third of its original topsoil in the past 150 years. Iowa has been particularly hard hit, already having lost half of its topsoil since the arrival of the first European settlers. In California, the current topsoil is being eroded at a rate that exceeds its replenishment rate by a factor of more than 70.

The amount of topsoil lost in a day is staggering: Emiliani estimates that the lost topsoil would fill a line of dump trucks 3,500 miles long! The cost of this loss of natural resource is incalculable. The map below summarizes points brought up in lecture on geographic distribution of soil degradation:

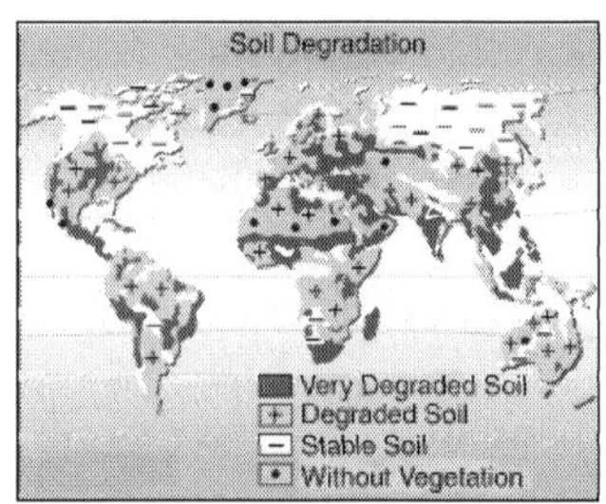

Soil is important for the support of life processes. Real soil only exists on Earth. Soil is the place where atmosphere, biosphere, hydrosphere, and lithosphere meet. Over 60,000 types of soils have been catalogued

Without soil, there is no food. Soils take 10's of thousands of years to form in rock weathering processes.

Weathering occurs due to a combination of chemical and mechanical processes that are subject to strong positive (reinforcing) feedback mechanisms. Over geologic time, rock weathering processes have led to sedimentary rocks and salty oceans!

Plants obtain inorganic minerals (nutrients) from the soil and incorporate their elements into biochemical materials. Animals may eat plants and each other, and synthesize new proteins, but the building blocks are the amino acids originally synthesized in plants. Soil effectively husbands the needed nutrients in the upper horizons.

The availability (or lack of availability) of important nutrients places critical constrains on the type of life that can survive in particular ecosystems. Soil erosion is a very serious problem globally because of its inextricable link to development activities. Soil conservation is therefore a very serious business and an important part of future global change studies.

SOIL ORGANISMS AND ORGANIC MATTER

Healthy soil teems with an immense community of living organisms. In fact, a hectare supports about 20,000 kilograms of soil organisms, equivalent to the weight of 40 horses. Although they make up only about 5 per cent of soil organic matter, these organisms are vital to many soil processes.

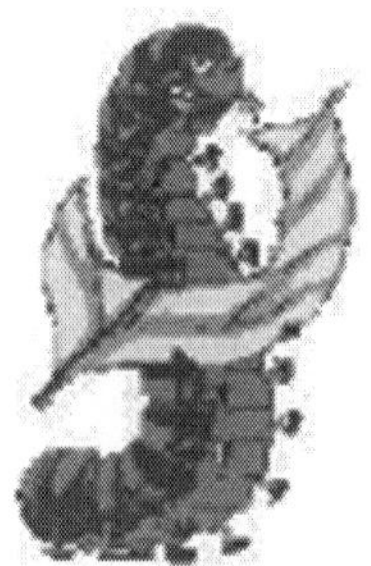

Through their roles in the decomposition cycle, they regulate the flow of energy through the soil, the cycling of nutrients, and the productivity of agroecosystems. Soil organisms span a wide range in size, from microscopic forms, such as bacteria, fungi, and protozoa, to large animals, such as insects, worms, and burrowing mammals.

The larger organisms assist in decomposition by ingesting plant residues, breaking them into finer particles, and mixing them as waste throughout the moist soil environment. These wastes become food for the microorganisms, which digest the organic matter, releasing plant nutrients and gases, and producing glues that stick the soil mineral particles together to form aggregates.

Macro soil organisms influence soils mainly by being mixers of soil materials. Ground squirrels, badgers, gophers (not the *Golden variety* Golden Spp) and crawfish are some animals that mix soil horizons with their digging.

Worms

Earthworms were described by Aristotle as 'the intestines of the earth'. Numerous soil scientists have been equally fascinated by the amount of work done by them, Charles Darwin said of them "It may be doubted whether there many other animals which have played so important a part in the history of the world as these lowly organized creatures".

Earthworms are miniature topsoil factories. The surface soil will eventually pass through an earthworm. Earth worms are amazingly strong, and can easily shift stones 60 times there own weight.

Worms may deposit 10 to 15 tons of castings per acre on the surface of the soil during a year. Research on the beneficial effects of worms generally does not show they will increase yields. However, we do know that the casts they leave behind are high in bacteria, organic matter, and plant nutrients. Castings have an NPK (nitrogen, phosphorus, potassium) ratio of 0.5-0.5-0.3 and are 50 per cent organic matter and 11 trace minerals. Worm castings work like time-released fertilizer.

Worms prefer a moist non-acid environment in which to live. They also need organic matter, which they use as a food source, and high amounts of available calcium. Worms also leave numerous channels in the soil which may result in pesticides and nutrients entering the subsoils at faster rates.

These channels allow preferential flow of water, rather than the water moving only through the soil pores.

All earthworms in Minnesota are exotics because the last period of glaciation wiped out all the natives at that time (if there were any). The earthworm appears to be rapidly altering the character of the sugar maple and basswood forests by consuming the leaf litter. The large deep burrowing night crawler seems to be responsible for most of the changes. Note the lack of the Oi and Oe layers in the soil with the abundant night crawlers.

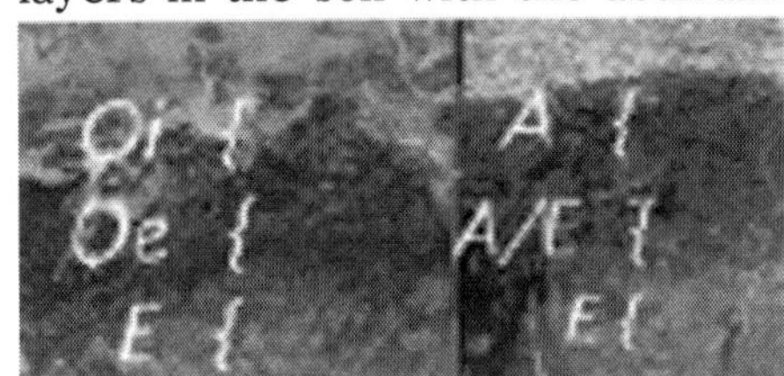

The non-native species have been introduced via bait containers and horticultural activities. Earthworms prefer the basic pH of the maple-basswood forest and are able to successfully exploit the food resources contained in the duff layers (O horizons). Normally a tree leaf may take three to five years to decompose and be incorporated into the soil humus. In forests infested with night crawlers, this process can take as little as four weeks. By accelerating the breakdown of plant material, earthworms change the way nutrients are recycled back to the plants. They may also be changing the ecology of the soil microorganism community be reducing the food for fungi and bacteria which rely on the duff layer as a main food source.

Worms should not be put into the soils of northern Minnesota to protect the ecology of the forest floor.

Roots

Roots absorb the water and nutrients that are needed by the plants for photosynthesis and respiration.

Roots in the soil play an important role in the activity of organisms. Roots are often a little leaky, and the material that they leak is referred to as root exudates. The area immediately around the root is known as the rhizosphere. The rhizosphere environment has a lower pH, and the soil atmosphere has lower O_2 and higher CO_2 concentrations. The rhizosphere is higher in soil organism activity due to increased food supply leaked by the root for the organisms to use. This includes: amino acids, organic acids, carbohydrates, nucleic acids, growth factors, enzymes, and soughed-off tissue. Benefits for the plant of having a rhizosphere include enhanced N mineralization, enhanced N_2 fixation, and nutrient solubilization.

The rhizosphere is defined as an intense zone of stimulated microbial activity around the root. Within the rhizosphere microbial numbers are much greater than in the bulk soil.

Microbes in the rhizosphere can be arbitrarily subdivided into the following groups: A. Pathogenic (invades and kills plants) B. Beneficial (often symbiotic with plants) C. Harmful (normally non-pathogenic opportunists on plants) D. Saprophytic (live on dead plants) E. Neutral (no effect on plants)

The microbes listed above are all competing for the some resources (space, nutrients and carbon) in the rhizosphere. The rhizosphere is a battlefield between pathogenic and non-pathogenic microorganisms. While we are mostly concerned with deleterious and pathogenic bacteria in the rhizosphere, there are some micro-organisms present in the rhizosphere which are good for roots.

An example of this is the actinomycete Streptomyces which secretes antibiotics and toxins into the soil which then inhibits the growth of other rhizosphere microorganisms. e.g. It prevents the spread of the pathogenic "damping-off" fungus.

Nematodes

Another worm-like organism is the nematode. Nematodes are microscopic worms that feed on organic matter and other soil animals or infect plant roots. Under a 10x hand lens, nematodes appear as transparent, thread-like worms. Parasitic nematodes are the most important from an agricultural standpoint. Many plants are affected, such as tomatoes, carrots, potatoes, peas, alfalfa, turfgrass, and fruit trees. Nematodes can parasitize virtually all crops and ornamental plants and can cause significant economic damage by reducing both yield and quality. Properly taken samples from small field units can reduce production costs by allowing the grower to eliminate nematodes

Lance nematodes, Hoplolaimus spp., are large nematodes which are highly resistant to effects of temperature extremes and dry soil conditions. One species, H. columbus, causes severe damage to soybeans and cotton. Another more widely distributed species, H. galeatus, is primarily a pathogen on grasses. Lance nematodes feed externally along root surfaces but may also feed with at least part of the body embedded in the root. Larvae look similar to adults except that they are smaller. This group of nematodes is easily detected with soil sampling. The life cycle of this nematode takes about 30 days under ideal conditions, and females lay eggs singly rather than in a mass. Though a female may lay as many as 100 eggs in a lifetime, that lifetime may last an entire growing season.

Root-knot nematodes, *Meloidogyne spp*., are one of the important plant-parasitic nematodes because of their wide host range and widespread distribution. Root-knot larvae enter roots of host plants near root tips and remain inside the root at one location throughout their life. As larvae feed, the root cells divide rapidly near the nematode's head. This rapid cell division and enlargement cause the swelling or knots on roots. .

Fungi

Fungi are not plants. Fungi were listed in the Plant Kingdom for many years. Then scientists learned that fungi show a closer relation to animals, but are unique and separate life forms. Now, Fungi are placed in their own Kingdom. The part of the fungus that we see is only the "fruit" of the organism. The living body of the fungus is a mycelium made out of a web of tiny filaments called hyphae. The mycelium is usually hidden in the soil, in wood, or another food source. A mycelium may fill a single ant, or cover many acres. The branching hyphae can add over a half mile (1 km) of total length to the mycelium each day. These webs live unseen until they develop mushrooms, puffballs, truffles, brackets, cups, "birds nests," "corals" or other fruiting bodies. If the mycelium produces microscopic fruiting bodies, people may never notice the fungus.

The most active decomposers of organic materials in a forested soil are the soil fungi. This is mainly because they are tolerant of acid soil conditions. All of us have seen fungi. Their size varies from single-cell yeasts to molds and mushrooms. The woody residue of the forest floor provides an abundance of food for certain fungi that are effective decomposers of lignin.

Most fungi build their cell walls out of chitin. This is the same material as the hard outer shells of insects and other arthropods. Plants do not make chitin. Fungi feed by absorbing nutrients from the organic material in which they live. Fungi do not have stomachs. They must digest their food before it can pass through the cell wall into the hyphae. Hyphae secrete acids and enzymes that break the surrounding organic material down into simple molecules they can easily absorb.

Mycorrhizae fungi and endomycorrhizae fungi

Mycorrhizae are fungi associated with the fine roots of most plants. The term itself means "fungus root". There are hundreds species of fungi which function as mycorrhiza; most are basidiomycetes, the class of fungi which form mushrooms. An individual plant may have several different mycorrhizae associated with its roots, and some mycorrhizae may be limited to only a few species of plant. These fungi can benefit plants by enhancing the nutrient absorbing ability of roots.

Mycorrhizae are especially important in facilitating uptake of phosphorous. This enhancement of nutrient uptake is a result of the extensive system of hyphae and mycelia (thread-like filaments of the mycorrhizal fungus) that pervade soils. They function like root hairs but are much more far reaching.

The relationship of this fungus with plants is a mutually beneficial one, with the fungi receiving energy in the form of carbohydrates from the host plant. There are two types of mycorrhizae. Filaments of the first type, called ectomycorrhizae, penetrate between cells of roots, but not into root cells, and also form a thick cylindrical sheath around young lateral roots. The affected roots become short and thickened, and are deficient in root hairs. It is believed that this sheath may protect roots from invasion by plant pathogens. Most trees have this type of mycorrhizae.

A second type, called endomycorrhizae, actually penetrate into root cells and extend out from roots like root hairs to absorb nutrients from the soil solution. They do not form a sheath around roots nor do they alter the structure of roots. Endomycorrhizae are found on a greater range of plants than are ectomycorrhizae.

Studies have repeatedly shown that plant growth is enhanced by the presence of mycorrhizae. Mycorrhizae are particularly abundant in forest soils but are found in almost all soils, with the possible exception of grasslands where no trees have previously grown. Growth enhancement is especially significant for plants growing on infertile soils and dry soils. Interestingly, mycorrhizae development decreases following heavy fertilization of soil. The reduced growth of the pine seedlings in the middle of this photo was because of the lack of mycorrhizae on the roots. These seedlings were planted in an old limestone rock roadbed. The soil has a pH greater than 8, which the fungus could not tolerate.

Bacteria

The most abundant organisms in the soil are bacteria. Bacteria are minuscule, one-celled organisms that can only be seen with a powerful light (100X) or electron microscope. They can be so numerous that a pinch of soil can contain millions of organisms. Soils often have between 1,000,000 to 10,000,000 bacteria per gram. Bacteria are tough, they occur everywhere on earth and have even been found over a mile down into the core of the earth.

Bacteria have an extremely varied metabolism. Bacteria can use reduced inorganics, the sun, or organics as an energy source. Some bacteria can live without free molecular oxygen.

Bacteria are common throughout the soil, but tend to be most abundant in or adjacent to plant roots, an important food source. Actinomycetes are a broad group of bacteria that form thread-like filaments in the soil. They are responsible for the distinctive scent of freshly exposed, moist soil. Actinomycetes are particularly effective at breaking down tough substances like cellulose (which makes up the cell walls of plants) and chitin (which makes up the cell walls of fungi) even under harsh conditions, such as high soil pH.

Free-living bacteria fix atmospheric nitrogen, adding it to the soil nitrogen pool. Other nitrogen-fixing bacteria form associations with the roots of leguminous plants such as lupine, clover, alfalfa, and milk vetch.

Actinomycetes form associations with some non-leguminous plants (important species are bitterbrush, mountain mahogany, cliff rose, and ceanothus) and fix nitrogen, which is then available to both the host and other plants in the near vicinity.

Some bacteria exude a sticky substance that helps bind soil particles into small aggregates. So despite their small size, they help improve water infiltration, water- holding capacity, soil stability, and aeration.

Bacteria are becoming increasingly important in bioremediation, meaning that we (people) can use bacteria to help us clean up our messes. Bacteria are capable of filtering and degrading a large variety of human-made pollutants in the soil and groundwater so that they are no longer toxic. The list of materials they can detoxify includes herbicides, heavy metals, and petroleum products.

Bacteria can be divided into 2 large groups based on their carbon source. Autotrophic bacteria are independent of any carbon in the soil since they fix atmospheric CO_2; and obtain energy from the reactions of nitrogen and sulphur compounds in the soil. Heterotrophic bacteria require carbon compounds as a food source. They are extremely important in decomposing organic matter. In the carbon cycle, CO_2; is absorbed by plants during photosynthesis. As these plants die and are incorporated into the soil, heterotrophic bacteria and other soil organisms decompose this organic matter and release CO_2; into the soil atmosphere. This movement of CO_2 into the atmosphere completes the carbon cycle.

The living organisms in the soil represent the key to plant health and to human health. Soil microorganisms are the essential link between mineral reserves and plant growth. *Ecological soil management* aims at assisting all soil organisms not substituting for them with a chemical system.

Organic material

"Organic material" or "organic matter" is informally used to denote a material that originated as a living organism; most such materials contain carbon and are capable of decay. They are found in soil and elsewhere, and may include still-living material, such a cell culture.

"Organic" matter is not necessarily created by living organisms, and living organisms are not made entirely of organic material. The more technically correct term for any material built by organismal material is "biotic." A clam's shell, for example, is an essential component of the living organism, but it is not technically organic. Conversely, urea is one of many organic substances that can be synthesized without any biological activity. The equation of "organic" with living organisms comes from the scientifically abandoned idea of vitalism that attributed a special force to life that alone could create organic substances, which was first called into question by the *a*biotic synthesis of urea by Friedrich Wöhler in 1828.

Examples of organic materials are wood, linoleum, straw, humus, manure, bark, crude oil, and cotton.

The use of organic materials is high on the agenda of many popular environmental groups because such materials are usually biodegradable, renewable, and the processing is commonly understood and has minimal environmental impact. However, not all organic materials can be considered environmentally friendly, such as fossil fuels, and other highly processed organic materials.

Processed organic material may be called bio-based material.

Bio-based material

A bio-based material is simply an engineering material made from substances derived from living tissues. These materials are sometimes referred to as biomaterials, but this word also has another meaning.

Strictly the definition could include many common materials such as wood and leather, but it typically refers to modern materials that have undergone more extensive processing.

Examples include:

Polylactic acid - A polymer produced by fermentation

Bioplastics - Including a soy oil based plastic now being used to make body panels for John Deere tractors

Zein- A natural biopolymer which is the most abundant corn protein

Cornstarch - The starch of the maize grain, used to create based packing pellets

Uses

Examples of the use of organic materials include:

building material, for a stylistic reasons, or to reduce allergic reactions.

energy production

clothing

composting and mulch

Biomass

Biomass is organic non-fossil material, collectively. In other words, 'biomass' describes the mass of all biological organisms, dead or alive, excluding biological mass that has been transformed by geological processes into substances such as coal or petroleum.

Typically, biomass refers to plant matter that can be processed into various industrial material, including biofuel (including biodiesel, methanol fuel, and ethanol fuel), and building materials, but can also include cellulose fibres used for biodegradable plastics and paper, among other things. Biomass can come from several plants, including switchgrass, hemp, corn, and sugarcane. Hemp is the best plant for biomass, although it is underused, and illegal in America. Biomass represents an environmental issue, because biomass is renewable and good for the environment. Biomass would become a huge industry if we moved from a hydrocarbon based economy (fossil fuel), to a carbohydrate (cellulose) economy.

Literally though, biomass refers to the cumulation of living matter. The most successful animal of the earth, in terms of biomass, is the Antarctic krill, *Euphausia superba*, with a biomass of probably over 500 million tons, roughly twice the total biomass of humans. The entire earth contains about 75 billion tons of biomass. Humans comprise about 250 million tons (0.33 per cent), domesticated animals about 700 million (1.0 per cent), and crops about 2 billion tons or 2.7 per cent of the Earth's biomass.

Biomass is also the dried organic mass of an ecosystem. As the trophic level increases, the biomass of each trophic level decreases. That is, producers (grass, trees, scrubs, etc.) will have a much higher biomass than animals that consume the producers (deer, zebras, insects, etc.). The level with the least biomass will be the highest predators in the food chain (foxes, eagles, etc.)

Detritus

In biology, detritus is organic waste material from decomposing dead plants or animals. This detritus is an important source of nutrients for so-called "detritus feeders" or detritivores. In many ecosystems, these form an important part of the food chain. What is left behind by the detritivores is then further broken down and recycled by decomposers, such as bacteria and fungi.

Humus

Humus is a word actually used for two different things, which are both related to soil and thus get used interchangably.

First, in earth sciences "humus" is any organic matter which has reached a point of stability, where it will break down no further and might, if conditions do not change, remain essentially as it is for centuries, or millenia.

Second, in agriculture, "humus" is often used simply to mean mature compost, or natural compost extracted from a forest or other spontaneous source for use to amend soil.

The process of "humification" can occur naturally in soil, or in the production of compost. Chemically stable humus (first definition) is thought by some to be important to the fertility of soils in both a physical and chemical sense, though some agricultural experts advocate a greater focus on other aspects of nutrient delivery, instead. Physically, it helps the soil retain moisture, and encourages the formation of good soil structure. Chemically, it has many active sites which bind to ions of plant nutrients, making them more available. Humus is often described as the 'life-force' of the soil. Yet it is difficult to define humus in precise terms; it is a highly complex substance, the full nature of which is still not fully understood. Physically, humus can be differentiated from organic matter in that the latter is rough looking material, with coarse plant remains still visible, while once fully humified it become more uniform in appearance (a dark, spongy, jelly-like substance) and amorphous in structure. That is, it has no determinate shape, structure or character.

Plant remains (including those that have passed through an animal and are excreted as manure) contain organic compounds: sugars, starches, proteins, carbohydrates, lignins, waxes, resins and organic acids. The process of organic matter decay in the soil begins with the decomposition of sugars and starches from carbohydrates which break down easily as saprotrophs initially invade the dead plant, while the remaining cellulose breaks down more slowly. Proteins decompose into amino acids at a rate depending on carbon to nitrogen ratios. Organic acids break down rapidly, while fats, waxes, resins and lignins remain relatively unchanged for longer periods of time. The humus that is the end product of this process is thus a mixture of compounds and complex life chemicals of plant, animal, or microbial origin, which has many functions and benefits in the soil.

Benefits of Humus

The mineralisation process that converts raw organic matter to the relatively stable substance that is humus feeds the soil population of micro-organisms and other creatures, thus maintaining high and healthy levels of soil life.

Effective and stable humus are further sources of nutrients to microbes, the former providing a readily available supply while the latter acts as a more long-term storage reservoir.

Humification of dead plant material causes complex organic compounds to break down into simpler forms which are then made available to growing plants for uptake through their root systems.

Humus is a colloidal as substance, and increases the soil's cation exchange capacity, hence its ability to store nutrients on clay particles; thus while these nutrient cations are accessible to plants, they are held in the soil safe from leaching away by rain or irrigation.

Humus can hold the equivalent of 80-90 per cent of its weight in moisture, and therefore increases the soil's capacity to withstand drought conditions.

The biochemical structure of humus enables it to moderate – or buffer – excessive acid or alkaline soil conditions.

During the Humification process, microbes secrete sticky gums; these contribute to the crumb structure of the soil by holding particles together, allowing greater aeration of the soil. Toxic substances such as heavy metals, as well as excess nutrients, can be chelated (that is, bound to the complex organic molecules of humus) and prevented from entering the wider ecosystem.

The dark colour of humus (usually black or dark brown) helps to warm up cold soils in the spring.

Humification of leaf litter and formation of clay-humus complexes

Compost which is readily capable of further decomposition is sometimes referred to as effective or active humus, though again actual scientists would say that if it is not stable, it's not humus at all. This kind of compost is principally derived from sugars, starches, and proteins, and consists of simple organic (fulvic) acids. It is an excellent source of plant nutrients, but of little value regarding long-term soil structure and tilth. Stable (or passive) humus consisting of humic acids, or humins, on the other hand, are so highly insoluble (or tightly bound to clay particles that they cannot be penetrated by microbes) that they are greatly resistant to further decomposition. Thus they add few readily available nutrients to the soil, but play an essential part in providing its physical structure. Some very stable humus complexes have survived for thousands of years. Stable humus tends to originate from woodier plant materials, eg, cellulose and lignins.

Green Manure

Green manure crops or cover crops are plants sown to enrich the soil when plowed or tilled under at a later date. Organic matter is added to the soil as well as the potential for nutrients such as nitrogen. An added benefit is reducing soil erosion during winter months due to blowing winds.

Legumes used as green manure crops can provide a significant amount of nitrogen to soil the following year. Typically, green manure crops are sown in the fall and rototilled under the following spring. However, fallow ground can be covered with plant material for a year or more before plowing under. Crops should be chosen carefully to avoid a potential weed problem down the road. While the crop can be plowed under, it may set seed or reproduce through root cuttings.

Common winter green manure crops include: rye, wheat and ryegrass. Most are sown between August and September to allow for fall growth. Summer cover crops include oats, soybeans and buckwheat. Other plants used for green manure crop include alfalfa, clover (red, alsike, crimson, sweet), barley, bromegrass, lespedeza. Most are sown in the early spring and plowed under in the fall.

C to N Ratio of Organic Materials and Composting

If organic material is added to soil that has a wide carbon to nitrogen ratio, the nitrogen in the soil will be used by the organisms to decompose the organic matter. This can lead to nitrogen deficiencies for plants growing in the soil. In order to avoid N immobilization it is best to compost organic materials that have a C:N ration wider than 30:1 before they are added to the soil.

Everything organic has a ratio of carbon to nitrogen (C:N) in its tissues.
Manure (Fresh) C: N=15:1
Legumes (peas etc.) 15:1
Grass Clippings 20:1
Manure w/Weeds 23:1
Weeds (Fresh) 25:1
Hay (Dry) 40:1
Leaves (Fresh) 40:1
Leaves (Dry) 60:1
Weeds (Dry) 90:1
Straw, cornstalks 100:1
Pine Needles 110:1
Sawdust 500:1
Wood Chips 700:1

It is the combination of materials that creates the ideal climate for compost microbes which is around a C:N ratio of 30:1. This combination, along with moisture, oxygen and surface area, is what makes a fast, hot pile.

Some composters like to keep things simple and use the terms brown (carbon) and green (nitrogen), and follow the general rule of 1 part brown for every 2 parts green. Other composters like to build their piles using a variety of materials and using the following formula to determine the C:N ratio of their compost pile. This balance is difficult to achieve exactly, but you can come close by combining materials and calculating the resulting (C:N) ratio.

For example, mixing two parts green grass clippings (which have a C:N ratio of 20:1) with one part dry leaves will give you a C:N ratio of 33:1, very close to the ideal.

The process is this: add 20/1 + 20/1 (the 2 parts green) with 60/1 (the leaves).

The total is 100/3 or 33/1. Another sample: Making a pile of 2 parts green grass, 1 part pine needles and 1 part fresh manure yields a C:N ratio of _____?

20/1 + 20/1 + 110/1 + 15/1=165/4=41/1 or this would need some additional "green" material.

Native SOM (soil organic matter) Levels

Under a forest ecosystem, organic matter will be maintained at high levels. However, when compared to grassland soils, the total quantity of organic matter in the soil is much lower. In a forest, organic matter is only added to the surface of the soil once each year when the leaves fall. It will take 1 to 3 years for the leaves to decompose and be added to the soil organic matter pool. It will take time to redistribute this organic matter into the soil. This will be aided by earthworms and other organisms.

Grasses have a fibrous root system that extensively explores the upper 18" of soil. During the year roots continually contribute fresh organic matter to the soil root zone. Thus the soil organic matter pool is throughout the fibrous root zone and not just as the surface as in the forest.

Climate has an influence on the quantity of organic residue. With increasing precipitation from areas of low rainfall to areas of high rainfall there is an accompanying increase in the amount of vegetation produced annually. Thus, the organic matter content of the soil increases from west to east across the central U.S.A... However, the difference between prairie and forest vegetation complicates this interpretation. Also, an increase in soil temperature speeds up the decomposition of organic debris and results in less OM in the soil.

Organic Geochemistry

Organic geochemistry is the study of the impacts and processes that organisms, and once-living organisms have on the earth.

Organic geochemistry includes studies of recent sediments to understand carbon cycling, climate change, and ocean processes, e.g.), and studies of ancient sediments, to understand the origins and sources of oil petroleum geochemistry.

PROBLEM SOILS

Introduction

Problem Soils impose a severe limitation on successful crop production due to the unfavourable effects of certain chemical and/or physical properties of soils. acid sulphate soils, saline and alkali soils, peat soils, soils with nutrient toxicity (very minor area) and nutrient deficiency are examples of chemical soil problems while steeply sloping soils, coarse textured soils, shallow soils, poorly drained soils, heavy textured soils and soils with ploughpan are included in soils with physical problems.

Chemical problems Acid Sulphate Soil

These soils occur near the coast on the Chittagong Coastal Plains and in the southwest of the Ganges Tidal Floodplain areas where the mangrove forest has been cleared for cultivation. These soils are also called Kosh soils in Chittagong. High acidity (pH 3.2-4.0) in surface soils, salinity (14-34 decisiemens per metre - dS/m), Al toxicity (8-103 ppm), low Zn and exchangeable Ca status, adverse physical and biological conditions with medium organic matter and poor tilth and activity of microorganisms limit crop growth. Embanking to prevent intrusion of saline or brackish water, improvement of drainage, applying lime to increase soil pH and flushing the embanked area with brackish water will reclaim these soils and make them suitable for producing more than one crop (at present only one crop is grown). However, the methods are costly and sometimes may not be economic also.

Saline soils For saline soils the electrical conductivity (EC) of a saturated extract of the soil is more than 4 dS/m. The exchangeable sodium percentage is less than about 15 and the pH usually is less than 8.5 because the salts are neutral. The sodium adsorption ratio (SAR) is less than 13.

Saline soils occur mainly in the Ganges Tidal Floodplain, Young Meghna Estuarine Floodplain and in the Tidal Floodplain areas of the Chittagong Coastal Plains and offshore islands.

High salinity, low fertility with respect to organic matter, nitrogen, zinc and copper, scarcity of quality irrigation water during winter, variability in rainfall from year to year, short winter season, heavy texture (silty clay to clay) and perennial waterlogging due to inadequate drainage hamper crop growth. Cultivation of salt tolerant crops, mulching in the summer, improvement of drainage and prevention of intrusion of saline water from sea will improve these soils.

Alkali soils In alkali soils, the exchangeable sodium percentage (ESP) is greater than 15, the electrical conductivity (EC) of a saturated extract is less than 4 ds/m, the pH is commonly 8.5 or less because of the presence of neutral salts and the sodium adsorption ratio (SAR) is at least 13.

Soils with dense and alkaline topsoils (pH >8.5) occupy very small areas among calcareous loamy ridge soils in widely scattered areas in the west of the Ganges River Floodplain. Topsoil alkalinity is the main problem. Breaking up the topsoil followed by cultivation of strong rooting grass for several years and application of sufficient organic manure will make the soil suitable for crop growth.

Peat soils The soils occur in the Gopalganj-Khulna Beels, deep depressions in the Sylhet Basin and in the Northern and Eastern Hills. The bulk density of peat surface soil is only 0.20-0.30 gm/cm3. Low bearing capacity when wet, strong acidity, low nutrient status and perennial wetness are constraints for crop production. Drying the peat soils eventually leads to shrinkage and acidification. Control of drainage allowing the soil not to dry, making raised cultivation beds and embankment may make the soils suitable for vegetable crops.

Phosphate fixing soils These soils include some Terrace soils developed on the Madhupur and barind tracts. Strongly acid soils rich in iron and aluminium and calcareous soils rich in calcium are prone to phosphate fixation. Appropriate methods of placing phosphatic fertilisers, addition of farmyard manure or compost and foliar application of phosphatic fertilisers will reduce harmful effects.

Iron, Zinc and Sulphur deficiency Iron and Zinc deficiencies occur only locally but may be widespread on the calcareous Ganges River Floodplain and sandy soils, peat soils, high pH saline soils and light textured Piedmont soils. Sulphur deficiencies in Bangladesh are acute and widespread in light textured and irrigated soils, where high yielding varieties are cultivated. The main causes are (a) increased use of high analysis sulphur free fertilisers like urea, TSP, (b) increased crop yield through HYV and (c) cultivation and continuous wetting of the soil. Application of zinc and sulphur containing fertilisers can easily eradicate zinc and sulphur problems.

Physical problems Steeply-sloping soils

These soils occur widely in hill areas and in a small proportion of side valleys in the Barind and madhupur tracts. Unfavourable slopes (quite often steep to very steep), shallowness of the soils profile, severe droughtiness in the dry season, moderate to rapid permeability and susceptibility to serious erosion pose problems to crop cultivation. Forest trees are best suited for very steep to steeply sloping soils. tea, Coffee, Rubber, banana etc on low hills with gentle slopes and deep loamy soils, and jackfruits in similar soils are possible options.

Attempts to increase organic matter, cultivation of cover crops wherever possible and leaving the soil surface barren as little as possible are steps to be considered for preserving the soil from any further destruction. Any alternative option would lead to catastrophe.

Coarse textured soils Lands with coarse textured soils mainly occur on the flood free Piedmont Plain, Active Floodplain, the hills and river chars. Low structural stability, low moisture holding capacity, low organic matter, zinc and sulphur limit crop production. Drought tolerant crops including tree crops on soils of high land, cultivation of fruits, spices or vegetables with mulching, controlled irrigation, farmyard manure and slow release fertilisers will be the best landuse.

Shallow soils Land with shallow soils (less than 50 cm deep) occurs in eroded part of the summits of the hills. Low fertility, low moisture status especially in dry season, insufficient space for proper root development and vulnerability to erosion limit crop production. Some parts are used for the cultivation of rice and others for dryland crops. Such lands should be kept under shrub, grass and dwarf bamboo in order to avoid soil degradation.

Poor drainage Soils with poor drainage occur in most of the active and meander floodplains and basins. Poor drainage develops unfavourable physical, chemical and biological conditions. Soils become devoid of oxygen, with consequent evolution of other gases such as carbon dioxide, methane and nitrogen, which affects upland crop production adversely. These soils are, however, suitable for rice cultivation and need drainage improvement for upland crops.

Soils with ploughpan Ploughpan is a dense soil layer occurring below the cultivated topsoil caused by pressure from the sole of the plough. Ploughpan is found in most soils, specifically in silt loam soils, which are cultivated in a wet condition and are puddled, for transplanted rice cultivation. It is advantageous for rice cultivation but provide limitations for upland crops. For cultivation of upland crops, ploughpans should be destroyed by using a deep plough. However, for cultivation of rice and upland crops in the same field in rotation, the ploughpan could be developed at a deeper depth using deep ploughing equipment.

In conclusion, low fertility (Northern and Eastern Hill soils, and Terrace soils), strong acidity (Piedmont Apron, Northern and Eastern Piedmonts, Basin Soils in dry season and Acid sulphate soils), droughtiness (Basin soils in dry season, and Northern and Eastern Piedmont Plains), wetness (Peat Soils, Hill Soils and Basin Soils in the early dry season), zinc deficiency (Basin Soils, Peat Soils, and Estuarine Floodplain Soils), phosphate fixation (Terrace soils), ploughpan (Tista Floodplain, Old Brahmaputra Floodplain, Barind Tract, Meghna Estuarine Floodplain), high permeability (Ganges Floodplain Soil, and Friable Red Clay Terrace Soils), low moisture status (Ganges Floodplain Soils, Barind Tract and Northern and Eastern Piedmont Soils) and heavy consistence (Terrace soils) and low bearing capacity are the major soil constraints for crop production in Bangladesh. Though the constraints referred to here are not mutually exclusive in their occurrence, the total area is quite significant and needs careful study so that a complete or partial remedy may be found, as Bangladesh is a densely populated country and is barely self-sufficient in grain production.

Water Management for Salinity Control

Introduction

Water management may mean: management of water treatment of drinking water, industrial water, sewage or wastewater management of water resources management of flood protection .

Water treatment Water treatment in a general sense refers to the treatement of water to make it more acceptable for what will be done with it (either usage or discharge into the environment). This may involve the removal of impurities as well as the use of chemicals to counter the effects of chemicals or organisms that cannot be completely removed (e.g., chlorine to kill germs). Such chemicals may either be just used in the treatement process or subsequently removed.

Water purification Water purification is the removal of contaminants from untreated water to produce drinking water that is pure enough for human consumption. Substances that are removed during the process of drinking water treatment include bacteria, algae, viruses, fungi, and man-made chemical pollutants. Many contaminants, such as man-made chemicals and heavy metals, can be dangerous—but depending on the quality desired, some are removed to improve the water's smell, taste, and appearance. There really is no such thing as pure water. As the universal solvent, the moment that purified water is exposed to the environment it interacts, even with carbon dioxide in the air. Water purification therefore is a process describing the treatments employed to meet the objectives of the user.

Sewage treatment Sewage treatment is the process that removes the majority of the contaminants from wastewater or sewage and produces both a liquid effluent suitable for disposal to the natural environment and a sludge. To be effective, sewage must be conveyed to a treatment plant by appropriate pipes and infrastructure and the process itself must be subject to regulation and controls. Some wastewaters require different and sometimes specialised treatment methods.

At the simplest level, treatment of sewage and most wastewaters is through separation of solids from liquids, usually by settlement. By progressively converting dissolved material into solids, usually a biological floc which is then settled out, an effluent stream of increasing purity is produced.

Sewage (or domestic wastewater) treatment incorporates physical, chemical and biological processes which treat and remove physical, chemical and biological contaminants from water following human use. The objective of the treatment is to produce both a clean wastestream (or treated effluent) suitable for discharge or reuse back into the environment, and a solid waste or sludge also suitable for proper disposal or reuse.

Sewage is generated by residences, institutions, and commercial and industrial establishments. It can be treated onsite at the point of which it is generated (e.g., septic tanks or onsite package plants), or collected and conveyed via a network of pipes and pump stations to a municipal treatment plant . Efforts to collect, treat and discharge domestic wastewater are typically subject to local, state and federal regulations and standards (regulation and controls). Industrial sources of wastewater often require specialized treatment processes .

Typically, sewage treatment is achieved by the initial physical separation of solids from the raw wastewater stream followed by the progressive conversion of dissolved biological matter into a solid biological mass using indigenous, water-borne bacteria. Once the biological mass is separated or removed, the treated water may undergo additional disinfection via chemical or physical processes. This 'final effluent' can then be discharged or re-introduced back into a natural surface water body (stream, river or bay) or other environment (wetlands, golf courses, greenways, etc.). The segregated biological solids undergo additional treatment and neutralization prior to proper disposal or re-use.

These treatment processes are typically referred to as

primary treatment (solids settlement), secondary treatment (biological treatment of the supernatant and settled solids) tertiary treatment (additional polishing stages such as lagooning, micro-filtration or disinfection).

Sewage is the liquid waste from toilets, baths, showers, kitchens, etc. that is disposed of via sewers. In many areas sewage also includes some liquid waste from industry and commerce. Much sewage also includes some surface water from roofs or hard-standing areas. Municipal wastewater therefore includes residential, commercial, and industrial liquid waste discharges, and may include stormwater runoff.

Sewerage systems that transport liquid waste discharges and stormwater together to a common treatment facility are called combined sewer systems.

As rainfall runs over the surface of roofs and the ground, it may pick up various contaminants including soil particles (sediment), heavy metals, organic compounds, animal waste, and oil and grease. Some jurisdictions require stormwater to receive some level of treatment before being discharged to the environment. Examples of treatment processes used for stormwater include sedimentation basins, wetlands, and vortex separators (to remove coarse solids).

The site where the process is conducted is called a sewage treatment plant. The flow scheme of a sewage treatment plant is generally the same for all countries:

Mechanical treatment;

Influx (Influent)

Removal of large objects

Removal of sand and grit

Pre-precipitation

Biological treatment;

Oxidation bed (oxidizing bed) or Aerated systems

Post precipitation

Effluent

Treatment stages

Primary treatment

Primary treatment is to reduce oils, grease, fats, sand, grit, and coarse (settleable) solids. This step is done entirely with machinery, hence the name mechanical treatment.

Influx (influent) and removal of large objects In the mechanical treatment, the influx (influent) of sewage water is strained to remove all large objects that are deposited in the sewer system, such as rags, sticks, condoms, sanitary towels (sanitary napkins) or tampons, cans, fruit, etc. This is most commonly done using a manual or automated mechanically raked screen. This type of waste is removed because it can damage the sensitive equipment in the sewage treatment plant. **Sand and grit removal** This stage typically includes a sand or grit channel where the velocity of the incoming wastewater is carefully controlled to allow sand grit and stones to settle but still maintain the majority of the organic material within the flow. This equipment is called a detritor or sand catcher. Sand grit and stones need to be removed early in the process to avoid damage to pumps and other equipment in the remaining treatment stages. Sometimes there is a sand washer (grit classifier) followed by a conveyor that transports the sand to a container for disposal. The contents from the sand catcher may be fed into the incinerator in a sludge processing plant but in many cases the sand and grit is sent to a land-fill.

Screening or maceration This stage typically includes a sand or grit channel where the velocity of the incoming wastewater is carefully controlled to allow sand grit and stones to settle but still maintain the majority of the organic material within the flow. This equipment is called a detritor or sand catcher. Sand grit and stones need to be removed early in the process to avoid damage to pumps and other equipment in the remaining treatment stages. Sometimes there is a sand washer (grit classifier) followed by a conveyor that transports the sand to a container for disposal. The contents from the sand catcher may be fed into the incinerator in a sludge processing plant but in many cases the sand and grit is sent to a land-fill.

Sedimentation Almost all plants have a sedimentation stage where the sewage is allowed to pass through large circular or rectangular tanks. The tanks are large enough that faecal solids can settle and floating material such as grease and plastics can rise to the surface and be skimmed off. The main purpose of the primary stage is to produce a generally homogeneous liquid capable of being treated biologically and a sludge that can be separately treated or processed. Primary settlement tanks are usually equipped with mechanically driven scrapers that continually drive the collected sludge towards a hopper in the base of the tank from where it can be pumped to further sludge treatment stages.

Secondary treatment Secondary treatment is designed to substantially degrade the biological content of the sewage such as are derived from human waste, food waste, soaps and detergent. The majority of municipal and industrial plants treat the settled sewage liquor using aerobic biological processes. For this to be effective, the biota require both oxygen and a substrate on which to live. There are number of ways in which this is done. In all these methods, the bacteria and protozoa consume biodegradable soluble organic contaminants (e.g. sugars, fats, organic short-chain carbon molecules, etc.) and bind much of the less soluble fractions into floc particles. Secondary treatment systems are classified as fixed film or suspended growth. In fixed film systems - such as rock filters - the biomass grows on media and the sewage passes over its surface. In suspended growth systems

- such as activated sludge - the biomass is well mixed with the sewage. Typically, fixed film systems require smaller footprints than for an equivalent suspended growth system; however, suspended growth systems are more able to cope with shocks in biological loading and provide higher removal rates for BOD and suspended solids than fixed film systems.

Roughing filters Roughing filters are intended to treat particularly strong or variable organic loads, typically industrial, to allow them to then be treated by conventional secondary treatment processes. They are typically tall, circular filters filled with open synthetic filter media to which sewage is applied at a relatively high rate. The design of the filters allows high hydraulic loading and a high flow-through of air. On larger installations, air is forced through the media using blowers. The resultant liquor is usually within the normal range for conventional treatment processes.

Activated sludge Activated sludge plants use a variety of mechanisms and processes to use dissolved oxygen to generate a biological floc that substantially removes organic material. It also traps particulate material and can, under ideal conditions, convert ammonia to nitrite and nitrate and ultimately to nitrogen gas.

Filter Beds (Oxidising beds) In older plants and plants receiving more variable loads, trickling filter beds are used where the settled sewage liquor is spread onto the surface of a deep bed made up of coke (carbonised coal), limestone chips or specially fabricated plastic media. Such media must have high surface areas to support the biofilms that form. The liquor is distributed through perforated rotating arms radiating from a central pivot. The distributed liquor trickles through this bed and is collected in drains at the base. These drains also provide a source of air which percolates up through the bed, keeping it aerobic. Biological films of bacteria, protozoa and fungi form on the medias' surfaces and eat or otherwise reduce the organic content. This biofilm is often fed on by insects and worms, which attract birds, which attract ornithologists.

Rotating plates and spirals In some smaller plants slowly revolving plates or spirals are used which are partially submerged in the liquor. A biotic floc is created which provides the required substrate.

Moving Bed Biological Reactor Moving Bed Biological Reactor (MBBR) involve the addition of inert media into existing activated sludge basins to provide active sites for biomass attachment. This conversion results in a strictly attached growth system. Advantages of attached growth systems include 1) maintain a high density of biomass population 2) increase the efficiency of the system without the need for increasing the mixed liquor suspended solids (MLSS) concentration and 3) eliminate the cost of operating the return activated sludge (RAS) line.

Biological Aerated Filters Biological Aerated (or Anoxic) Filter (BAF) combines filtration with biological carbon reduction, nitrification or denitrification. BAF usually includes a reactor filled with a filter media. The media is either in suspension or supported by a gravel layer at the foot of the filter. The dual purpose of this media is to support highly active biomass that is attached to it and to filter suspended solids. Carbon reduction and ammonia conversion occurs in aerobic mode and sometime achieved in a single reactor while nitrate conversion occurs in anoxic mode. BAF is operated either in upflow or downflow configuration depending on design specified by manufacturer.

Membrane Biological Reactors Membrane Biological Reactors (MBR) includes a semi-permeable membrane barrier system either submerged or in conjunction with an activated sludge process. This technology guarantees removal of all suspended and some dissolved pollutants. The limitation of MBR systems is directly proportional to nutrient reduction efficiency of the activated sludge process. The cost of building and operating a MBR is usually higher than conventional wastewater treatment.

Secondary sedimentation The final step in the secondary treatment stage is to settle out the biological floc or filter material and produce sewage water containing very low levels of organic material and suspended matter.

Tertiary treatment Tertiary treatment provides a final stage to raise the effluent quality to the standard required before it is discharged to the receiving environment (sea, river, lake, ground, etc.) More than one tertiary treatment process may be used at any treatment plant. If disinfection is practiced, it is always the final process. It is also called Effluent polishing.

Filtration Sand filtration removes much of the residual suspended matter. Filtration over activated carbon removes residual toxins.

Lagooning Lagooning provides settlement and further biological improvement through storage in large man-made ponds or lagoons. These lagoons are highly aerobic and colonization by native macrophytes, especially reeds, is often encouraged. Small filter feeding invertebrates such as Daphnia and species of Rotifera greatly assist in treatment by removing fine particulates.

Constructed wetlands Constructed wetlands include engineered re-edbeds and a range of similar methodologies, all of which provide a high degree of aerobic biological improvement and can often be used instead of secondary treatment for small communities.

Nutrient removal Wastewater may also contain high levels of nutrients (nitrogen and phosphorus) that in certain forms may be toxic to fish and invertebrates at very low concentrations (e.g. ammonia) or that can create nuisance conditions in the receiving environment (e.g. weed or algal growth). Weeds and algae may seem to be an aesthetic issue, but algae can produce toxins, and their death and consumption by bacteria (decay) can deplete oxygen in the water and suffocate fish and other aquatic life. Where receiving rivers discharge to lakes or shallow seas, the added nutrients can cause severe eutrophication losing many sensitive clean water fish. The removal of nitrogen and/or phosphorus from wastewater can be achieved either biologically or by chemical precipitation.

Nitrogen removal is effected through the biological reduction of nitrogen from the ammonia to nitrate (nitrification involving nitrifying bacteria such as Nitrobacter and Nitrosomous), and then from nitrate to nitrogen gas (denitrification), which is released to the atmosphere. These conversions require carefully controlled conditions to encourage the appropriate biological communities to form. Sand filters, lagooning and reed beds can all be used to reduce nitrogen. Sometimes the conversion of toxic ammonia to nitrate alone is referred to as tertiary treatment.

Phosphorus removal can be effected biologically in a process called enhanced biological phosphorus removal. In this process specific bacteria, called Polyphosphate accumulating Organisms, are selectively enriched and accumulate large quantities of phosphorus within their cells. When the biomass enriched in these bacteria is separated from the treated water, the bacterial biosolids have a high fertilizer value. Phosphorus removal can also be achieved, usually by chemical precipitation with salts of iron (e.g. ferric chloride) or aluminum (e.g. alum). The resulting chemical sludge, however, is difficult to dispose of, and the use of chemicals in the treatment process is expensive. Although this makes operation difficult and often messy, chemical phosphorous removal requires significantly smaller equipment footprint than biological removal and is easier to operate.

Disinfection The purpose of disinfection in the treatment of wastewater is to substantially reduce the number of living organisms in the water to be discharged back into the environment. The effectiveness of disinfection depends on the quality of the water being treated (e.g., turbidity, pH, etc.), the type of disinfection being used, the disinfectant dosage (concentration and time), and other environmental variables. Turbid water will be treated less successfully since solid matter can shield organisms, especially from Ultraviolet light or if contact times are low. Generally, short contact times, low doses and high flows all militate against effective disinfection. Common methods of disinfection include ozone, chlorine, or UV light. Chloramine, which is used for drinking water, is not used in waste water treatment because of its persistence.

Chlorination remains the most common form of wastewater disinfection in North America due to its low cost and long-term history of effectiveness. One disadvantage is that chlorination of residual organic material can generate chlorinated-organic compounds that may be carcinogenic or harmful to the environment. Residual chlorine or chloramines may also be capable of chlorinating organic material in the natural aquatic environment. Further, because residual chlorine is toxic to aquatic species, the treated effluent must also be chemically dechlorinated, adding to the complexity and cost of treatment. Ultraviolet (UV) Light is becoming the most common means of disinfection in the UK because of the concerns about the impacts of chlorine in chlorinating residual organics in the wastewater and in chlorinating organics in the receiving water. UV radiation is used to damage the genetic structure of bacteria, viruses, and other pathogens, making them incapable of reproduction. The key disadvantages of UV disinfection are the need for frequent lamp maintenance and replacement and the need for a highly treated effluent to ensure that the target microorganisms are not shielded from the UV radiation (i.e., any solids present in the treated effluent may protect microorganisms from the UV light).

Ozone O_3 is generated by passing oxygen O_2 through a high voltage potential resulting in a third oxygen atom becoming attached and forming O_3. Ozone is very unstable and reactive and oxidizes most organic material it comes in contact with, thereby destroying many disease-causing microorganisms. Ozone is considered to be safer than chlorine because, unlike chlorine which has to be stored on site (highly poisonous in the event of an accidental release), ozone is generated onsite as needed. Ozonation also produces fewer disinfection by-products than chlorination. A disadvantage of ozone disinfection is the high cost of the ozone generation equipment and the requirements for highly skilled operators.

Package plants and batch reactors In order to use less space, treat difficult waste, deal with intermittent flow or achieve higher environmental standards, a number of designs of hybrid treatment plants have been produced. Such plants often combine all or at least two stages of the three main treatment stages into one combined stage. In the UK, where a large number of sewage treatment plants serve small populations, package plants are a viable alternative to building discrete structures for each process stage.

For example, one process which combines secondary treatment and settlement is the Sequential Batch Reactor (SBR). Typically, activated sludge is mixed with raw incoming sewage and mixed and aerated. The resultant mixture is then allowed to settle producing a high quality effluent. The settled sludge is run off and re-aerated before a proportion is returned to the head of the works. SBR plants are now being deployed in many parts of the world including North Liberty, Iowa, and Llanasa, North Wales.

The disadvantage of such processes is that precise control of timing, mixing and aeration is required. This precision is usually achieved by computer controls linked to many sensors in the plant. Such a complex, fragile system is unsuited to places where such controls may be unreliable, or poorly maintained, or where the power supply may be intermittent.

Package plants may be referred to as *high charged* or *low charged*. This refers to the way the biological load is processed. In high charged systems, the biological stage is presented with a high organic load and the combined floc and organic material is then oxygenated for a few hours before being charged again with a new load. In the low charged system the biological stage contains a low organic load and is combined with floculate for a relatively long time.

Sludge treatment The coarse primary solids and secondary biosolids accumulated in a wastewater treatment process must be treated and disposed of in a safe and effective manner. This material is often inadvertently contaminated with toxic organic and inorganic compounds (e.g. heavy metals). The purpose of digestion is to reduce the amount of organic matter and the number of disease-causing microorganisms present in the solids. The most common treatment options include anaerobic digestion, aerobic digestion, and composting.

Anaerobic digestion Anaerobic digestion is a bacterial process that is carried out in the absence of oxygen. The process can either be thermophilic digestion in which sludge is fermented in tanks heated to about 38°C or mesophilic digestion where sludge is maintained in large tanks for weeks to allow natural mineralisation of the sludge. Thermophilic digestion generates biogas with a high proportion of methane that may be used to both heat the tank and run engines or microturbines for other on-site processes. In large treatment plants sufficient energy can be generated in this way to produce more electricity than the machines require. The methane generation is a key advantage of the anaerobic process. Its key disadvantage is the long time required for the process (up to 30 days) and the high capital cost.

The Goldbar Wastewater Treatment Plant in Edmonton, Alberta, Canada currently uses the process. Under laboratory conditions it is possible to directly generate useful amounts of electricity from organic sludge using naturally occurring electrochemically active bacteria. Potentially, this technique could lead to an ecologically positive form of power generation, but in order to be effective such a microbial fuel cell must maximize the contact area between the effluent and the bacteria-coated anode surface, which could severely hamper throughput.

Aerobic digestion Aerobic digestion is a bacterial process occurring in the presence of oxygen. Under aerobic conditions, bacteria rapidly consume organic matter and convert it into carbon dioxide. Once there is a lack of organic matter, bacteria die and are used as food by other bactieria. This stage of the process is known as *endogenous respiration*. Solids reduction occurs in this phase. Because the aerobic digestion occurs much faster than anaerobic digestion, the capital costs of aerobic digestion are lower. However, the operating costs are characteristically much greater for aerobic digestion because of energy costs for aeration needed to add oxygen to the process.

Composting Composting is also an aerobic process that involves mixing the wastewater solids with sources of carbon such as sawdust, straw or wood chips. In the presence of oxygen, bacteria digest both the wastewater solids and the added carbon source and, in doing so, produce a large amount of heat. Both anaerobic and aerobic digestion processes can result in the destruction of disease-causing microorganisms and parasites to a sufficient level to allow the resulting digested solids to be safely applied to land used as a soil amendment material (with similar benefits to peat) or used for agriculture as a fertilizer provided that levels of toxic constituents are sufficiently low.

Thermal depolymerization Thermal depolymerization uses hydrous pyrolysis to convert reduced complex organics to oil. The premacerated, grit-reduced sludge is heated to 250C and compressed to 40 MPa. The hydrogen in the water inserts itself between chemical bonds in natural polymers such as fats, proteins and cellulose. The oxygen of the water combines with carbon, hydrogen and metals. The result is oil, light combustible gases such as methane, propane and butane, water with soluble salts, carbon dioxide, and a small residue of inert insoluble material that resembles powdered rock and char. All organisms and many organic toxins are destroyed. Inorganic salts such as nitrates and phosphates remain in the water after treatment at sufficiently high levels that further treatment is required.

The energy from decompressing the material is recovered, and the process heat and pressure is usually powered from the light combustible gases. The oil is usually treated further to make a refined useful light grade of oil, such as no. 2 diesel and no. 4 heating oil, and then sold.

The choice of a wastewater solid treatment method depends on the amount of solids generated and other site-specific conditions. However, in general, composting is most often applied to smaller-scale applications followed by aerobic digestion and then lastly anaerobic digestion for the larger-scale municipal applications.

Sludge disposal When a liquid sludge is produced, further treatment may be required to make it suitable for final disposal. Typically, sludges are thickened (dewatered) to reduce the volumes transported off-site for disposal. Processes for reducing water content include lagooning in drying beds to produce a cake that can be applied to land or incinerated; pressing, where sludge is mechanically filtered, often through cloth screens to produce a firm cake; and centrifugation where the sludge is thickened by centrifugally separating the solid and liquid. Sludges can be disposed of by liquid injection to land or by disposal in a landfill. There are concerns about sludge incineration because of air pollutants in the emissions, along with the high cost of supplemental fuel, making this a less attractive and less commonly constructed means of sludge treatment and disposal. There is no process which completely eliminates the requirements for disposal of biosolids.

In South Australia, after centrifugation, the sludge is then completely dried by sunlight. The nutrient rich biosolids are then provided to farmers free-of-charge to use as a natural fertilizer. This method has reduced the amount of landfill generated by the process each year. **Treatment in the receiving environment** The outlet of a wastewater treating plant flows into a small river. Many processes in a wastewater treatment plant are designed to mimic the natural treatment processes that occur in the environment, whether that environment is a natural water body or the ground. If not overloaded, bacteria in the environment will consume organic contaminants, although this will reduce the levels of oxygen in the water and may significantly change the overall ecology of the receiving water. Native bacterial populations feed on the organic contaminants, and the numbers of disease-causing microorganisms are reduced by natural environmental conditions such as predation, exposure to ultraviolet radiation, etc. Consequently in cases where the receiving environment provides a high level of dilution, a high degree of wastewater treatment may not be required. However, recent evidence has demonstrated that very low levels of certain contaminants in wastewater, including hormones (from animal husbandry and residue from human birth control pills) and synthetic materials such as phthalates that mimic hormones in their action, can have an unpredictable adverse impact on the natural biota and potentially on humans if the water is re-used for drinking water. In the US and EU, uncontrolled discharges of wastewater to the environment are not permitted under law, and strict water quality requirements are to be met. A significant threat in the coming decades will be the increasing uncontrolled discharges of wastewater within rapidly developing countries.

Management and reclamation of salt-affected soils

Salt Sources Saline soils are found throughout Colorado. These salts originate from the natural weathering of minerals or from fossil salt deposits left from ancient sea beds. Salts accumulate in the soil of arid climates as irrigation water or groundwater seepage evaporates, leaving minerals behind. Irrigation water often contains salts picked up as water moves across the landscape, or the salts may come from human-induced sources such as municipal runoff or water treatment. As water is diverted in a basin, salt levels increase as the water is consumed by transpiration or evaporation.

Table. Common salt compounds. Salts are ionic crystalline compounds consisting of a cation and an anion.

Salt compound	Cation (+)	Anion (-)	Common name
$NaCl$	sodium	chloride	halite (table salt)
Na_2SO_4	sodium	sulfate	Glauber's salt
$MgSO_4$	magnesium	sulfate	epsom salts
$NaHCO_3$	sodium	bicarbonate	baking soda
Na_2CO_3	sodium	carbonate	sal soda
$CaSO_4$	calcium	sulfate	gypsum
$CaCO_3$	calcium	carbonate	calcite (lime)

Measuring Soil Salinity Saline soils contain large amounts of water soluble salts that inhibit seed germination and plant growth. The salts are white, chemically neutral, and include chlorides, sulfates, carbonates and sometimes nitrates of calcium, magnesium, sodium and potassium in Table above. Salinity is measured by passing an electrical current through a soil solution extracted from a saturated soil sample. The ability of the solution to carry a current is called electrical conductivity (EC). EC is measured in deciSiemens per meter (dS/m), which is the numerical equivalent to the old measure of millimhos per centimeter in Table below. The lower the salt content of the soil, the lower the dS/m rating and the less the effect on plant growth. Yields of most crops are not significantly affected where salt levels are 0 to 2 dS/m. Generally, a level of 2 to 4 dS/m affects some crops. Levels of 4 to 5 dS/m affect many crops and above 8 dS/m affect all but the very tolerant crops.

Table . Terms, units and conversions.

Symbol	Meaning	Units
Total Salinity		
TDS	Total dissolved solids	mg/L[a]; ppm[b]
EC	Electrical conductivity	dS/m[c]; mmho/cm[d]; μmho/cm[e]
Conversions		
1 dS/m = 1 mmho/cm = 1000 μmho/cm		
1 mg/L = 1 ppm		

[a]mg/L = milligrams per liter; [b]ppm = parts per million; [c]dS/m = deciSiemens per meter at 25° C; [d]mmho/cm = millimhos per centimeter at 25° C; [e]μmho/cm = micromhos per centimeter at 25°C

Saline soils cannot be reclaimed by chemical amendments, conditioners or fertilizers. A field can only be reclaimed by removing salts from the plant root zone. In some cases, selecting salt-tolerant crops may be needed in addition to managing soils.

There are three ways to manage saline soils. First, salts can be moved below the root zone by applying more water than the plant needs. This method is called the leaching requirement method. The second method, where soil moisture conditions dictate, combines the leaching requirement method with artificial drainage. Third, salts can be moved away from the root zone to locations in the soil, other than below the root zone, where they are not harmful. This third method is called managed accumulation.

Leaching Requirement For most surface irrigation systems in Colorado (furrow and flood), irrigation inefficiency (or over-irrigation) generally is adequate to satisfy the leaching requirement. However, poor irrigation uniformity often results in salt accumulation in parts of a field or bed. Surface irrigators should compare leaching requirement values to measurements of irrigation efficiency to determine if additional irrigation is needed. Adding more water to satisfy a leaching requirement reduces irrigation efficiency and may result in the loss of nutrients or pesticides and further dissolution of salts from the soil profile.

Leaching is accomplished on a limited basis at key times during the growing season, particularly when a grower may have high quality water available. Surface water in most areas of the state tend to have lower salinity than shallow, alluvial groundwater.

Deep groundwater may have an even lower salinity than either shallow groundwater or surface water. In situations where a grower has multiple water sources of varying quality, consider planned leaching events at key salinity stress periods for a given crop.

Most crops are highly sensitive to salinity stress in the germination and seedling stages. Once the crop growns past these stages, it can often tolerate and grow well in higher salinity conditions. Planned periodic leaching events might include a post-harvest irrigation to push salts below the root zone to prepare the soil (especially the seedbed/surface zone) for the following spring. Fall is the best time for a large, planned leaching event because nutrients have been drawn down. However, since each case is site-specific, examine the condition of the soil, groundwater, drainage, and irrigation system for a given field before developing a sound leaching plan.

Leaching Plus Artificial Drainage Where shallow water tables limit the use of leaching, artificial drainage may be needed. Cut drainage ditches in fields below the water table level to channel away drainage water and allow the salts to leach out. Drainage tile or plastic drainpipe can also be buried in fields for this purpose. Proper design and construction of a drainage system is critical and should be performed by a trained professional, such as your local USDA-Natural Resources Conservation Service (NRCS).

With all artificial drainage systems you must also consider disposal of the drainage water. Restrictions on the discharge of drain water to streams may apply in certain situations and should be investigated through the Colorado Department of Public Health and Environment. In the case of regulated discharge, treatment or collection and evaporation of the water on site may be required and may add significant costs.

The advantage of artificial drainage is that it provides the ability to use high quality, low salinity irrigation water (if available to a grower) to completely remove salts from the soil. However, artificial drainage systems will not work where there is no saturated condition in the soil. Water will not collect in a drain if the soil around it is not saturated.

Table. Estimated water application needed to leach salts.

Percent Salt Reduction	Amount of Water Required
50 %	6 inches
80%	12 inches
90%	24 inches

Example: If a soil's electrical conductivity is 8 mmhos/cm, and you want to reduce it to 4 mmhos/cm. This represents a 50 percent reduction in salts. Therefore, 6 inches of water would be required. After drainage appears adequate, the leaching process can begin. Table shows how much water is required to leach salts. Actual salt reduction depends upon water quality, soil texture and drainage.

Managed Accumulation In addition to leaching salt below the root zone, salts can also be moved to areas away from the primary root zone with certain crop bedding and surface irrigation systems. Figures 1 and 2 illustrate several ways to manage salt accumulation in this manner. The goal is to ensure the zones of salt accumulation stay away from germinating seeds and plant roots. Irrigation uniformity is essential with this method. Without uniform distribution of water, salts will build up in areas where the germinating seeds and seedling plants will experience growth reduction and possibly death.

Double-row bed systems require uniform wetting towards the middle of the bed. This leaves the sides and shoulders of the bed relatively free from injurious levels of salinity. Without uniform applications of water (one furrow receiving more or less than another), salts accumulate closer to one side of the bed. Periodic leaching of salts down from the soil surface and below the root zone may still be required to ensure the beds are not eventually salted out.

Alternate furrow irrigation may be desired for single-row bed systems. This is accomplished by irrigating every other furrow and leaving alternating furrows dry. Salts are pushed across the bed from the irrigated side of the furrow to the dry side.

Care is needed to ensure enough water is applied to wet all the way across the bed to prevent build up in the planted area. This method of salinity management can still result in plant injury if large amounts of natural rainfall fill the normally dry furrows and push salts back across the bed towards the plants. This phenomenon also occurs if the normally dry furrows are accidentally irrigated.

Sprinkler Irrigation Sprinkler-irrigated fields with poor water quality present a challenge because it is difficult to apply enough water to leach the salts and you cannot effectively utilize row or bed configurations to manage accumulation. Growers should monitor the soil EC and irrigation water salinity. Where adequate irrigation water exists above crop requirements, a leaching fraction (or percent of additional water needed above crop requirements) can be calculated for sprinkler irrigated fields using this equation:

$$\% \textit{Leaching requirement} = \frac{EC water}{2xEC\max} x100$$

In this equation, EC max is the maximum soil EC wanted in the root zone. Apply this leaching fraction to coincide with periods of low soil N and residual pesticide. Again, fall is an optimal time to move salts below the root zone.

Table. Potential yield reduction from saline soils for selected crops.

	Relative yield decrease %			
	0	10	25	50
Field crops	(EC_e)			
Barley	8.0	10.0	13.0	18.0
Sugarbeets*	7.0	8.7	11.0	15.0
Wheat	6.0	7.4	9.5	13.0
Sorghum	4.0	5.1	7.2	11.0
Soybean	5.0	5.5	6.2	7.5
Corn	1.7	2.5	3.8	5.9
Bean	1.0	1.5	2.3	3.3
Forages				
Tall wheatgrass	7.5	9.9	13.3	19.4
Wheatgrass	7.5	9.0	11.0	15.0
Crested wheatgrass	3.5	6.0	9.8	16.0
Tall fescue	3.9	5.8	8.6	13.3
Orchardgrass	1.5	3.1	5.5	9.6
Alfalfa	2.0	3.4	5.4	8.8
Meadow foxtail	1.5	2.5	4.1	6.7
Cloveralsike, red, ladino, strawberry	1.5	2.3	3.6	5.7
Bluegrass and other turf **				
Vegetables				
Broccoli	2.8	3.9	5.5	8.2
Cucumber	2.5	3.3	4.4	6.3
Cantaloupe	2.2	3.6	5.7	9.1
Spinach	2.0	3.3	5.3	8.6
Cabbage	1.8	2.8	4.4	7.0
Potato	1.7	2.5	3.8	5.9
Sweet corn	1.7	2.5	3.8	5.9
Lettuce	1.3	2.1	3.2	5.2
Onion	1.2	1.8	2.8	4.3
Carrot	1.0	1.7	2.8	4.6

*Sensitive during germination and emergence, EC_e should not exceed 3dS/m at this time.

Excerpted from R. S. Ayers and D.W. Westcot, 1976, Water Quality for Agriculture, Irrigation and Drainage Paper 29, FAO, Rome. Crop salt tolerance data in the table were developed, almost entirely, by the U.S. Salinity Laboratory, Riverside, CA.

Crop Tolerance to Soil Salinity Excessive soil salinity reduces the yield of many crops. This ranges from a slight crop loss to complete crop failure, depending on the type of crop and the severity of the salinity problem.

Although several treatments and management practices can reduce salt levels in the soil, there are some situations where it is either impossible or too costly to attain desirably low soil salinity levels. In some cases, the only viable management option is to plant salt-tolerant crops. Sensitive crops, such as pinto beans, cannot be managed profitably in saline soils. Table shows the relative salt tolerance of field, forage, and vegetable crops. The table shows the approximate soil salt content (expressed as the electrical conductivity of a saturated paste extract (ECe) in dS/m at 25 degrees C) where 0, 10, 25, and 50 percent yield decreases may be expected. Actual yield reductions will vary depending upon the crop variety and the climatic conditions during the growing season.

Fruit crops may show greater yield variation because a large number of rootstocks and varieties are available. Also, stage of plant growth has a bearing on salt tolerance. Plants are usually most sensitive to salt during the emergence and early seedling stages. Tolerance usually increases as the crop develops.

The salt tolerance values apply only from the late seedling stage through maturity, during the period of most rapid plant growth. Crops in each class are generally ranked in order of decreasing salt tolerance.

Other Management Options

Residue Management Crop residue at the soil surface reduces evaporative water losses, thereby limiting the upward movement of salt (from shallow, saline groundwater) into the root zone. Evaporation and thus, salt accumulation, tends to be greater in bare soils. Fields need to have 30 percent to 50 percent residue cover to significantly reduce evaporation. Under crop residue, soils remain wetter, allowing fall or winter precipitation to be more effective in leaching salts, particularly from the surface soil layers where damage to crop seedlings is most likely to occur.

Plastic mulches used with drip irrigation effectivly reduce salt concentration from evaporation. Sub-surface drip irrigation pushes salts to the edge of the soil wetting front, reducing harmful effects on seedlings and plant roots.

Pre-plant Irrigation As mentioned before, most crop plants are more susceptible to salt injury during germination or in the early seedling stages. An early-season application of good quality water, designed to fill the root zone and leach salts from the upper 6 to 12 inches of soil, may provide good enough conditions for the crop to grow through its most injury-prone stages

Irrigation Frequency Management Salts are most efficiently leached from the soil profile under higher frequency irrigation (shorter irrigation intervals). Keeping soil moisture levels higher between irrigation events effectively dilutes salt concentrations in the root zone, thereby reducing the salinity hazard.

Most surface irrigation systems (flood or furrow systems) cannot be controlled to apply less than 3 or 4 inches of water per application and are not generally suited to this method of salinity control. Sprinkler systems, particularly center-pivot and linear-move systems configured with low energy precision application (LEPA) nozzle packages or properly spaced drop nozzles, and drip irrigation systems provide the best control to allow this type of salinity management.

Soil Drainage Good drainage is complementary to irrigation. In those areas where there is plenty of irrigation water as in Punjab, Haryana nd Uttar Pradesh drainage has become almost a national problem and is agitating the minds of the farmers and of the government. Water accumulates in the soil either from rains or from seepage from canals. The high sub soil l water table damages agricultural land in many ways, leading to the accumulation of undesirable salts in the root zone or on the surface of the soil involving drainage problems.

The problem of persistant water logging is often complex and requires expert advice. In areas of heavy rainfall good surface drainage has to be arranged. The drainage ditches should be wide and of trapezodial or parabolic shape so that they offer a rapidly increasing cross section to the flow of the stream as the water level in them rises. On relatively flat land they are made to run down the slope. The velocity of water in the drains should be maintained at non erosive levels by proper grading and shaping. The spacing of the field drains should be worked out taking into consideration the character of the soil, the slope, the intensity of rainfall etc. A field drain should be laid to join a utility channel of adequate dimension. These utility channels should be sufficiently big to take the maximum quantity of water.

In those soils where the water accumulates underground within a depth of 0.6 to 1.2m underground drainage is needed. In case this high water table is due to percolation from an external source, intersection drains should be installed to catch and divert the water. There are several types of drainage systems but most of them are costly. A master plan for draining the land should be prepared in consultation with the specialist engineer and after making a preliminary survey of the levels.

The best kinds of drainage pipes are hard, burnt, earthenware flangeless pipes. The diameter of the pipe is usually less than10cm. The pipes should be laid below the lowest root level of the crops proposed to tbe grown on the land. The joints should be left uncemented but protected by a strip of tar paper to prevent the entry of soil and silt as the drainage takes place largely through the joints and not through the porous body as is beleived. *Example*: If a soil's electrical conductivity is 8 mmhos/cm, and you want to reduce it to 4 mmhos/cm. This represents a 50 percent reduction in salts. Therefore, 6 inches of water would be required. After drainage appears adequate, the leaching process can begin. Table shows how much water is required to leach salts. Actual salt reduction depends upon water quality, soil texture and drainage.

Managed Accumulation In addition to leaching salt below the root zone, salts can also be moved to areas away from the primary root zone with certain crop bedding and surface irrigation systems. Figures 1 and 2 illustrate several ways to manage salt accumulation in this manner. The goal is to ensure the zones of salt accumulation stay away from germinating seeds and plant roots. Irrigation uniformity is essential with this method. Without uniform distribution of water, salts will build up in areas where the germinating seeds and seedling plants will experience growth reduction and possibly death.

Double-row bed systems require uniform wetting towards the middle of the bed. This leaves the sides and shoulders of the bed relatively free from injurious levels of salinity. Without uniform applications of water (one furrow receiving more or less than another), salts accumulate closer to one side of the bed. Periodic leaching of salts down from the soil surface and below the root zone may still be required to ensure the beds are not eventually salted out.

Alternate furrow irrigation may be desired for single-row bed systems. This is accomplished by irrigating every other furrow and leaving alternating furrows dry. Salts are pushed across the bed from the irrigated side of the furrow to the dry side. Care is needed to ensure enough water is applied to wet all the way across the bed to prevent build up in the planted area. This method of salinity management can still result in plant injury if large amounts of natural rainfall fill the normally dry furrows and push salts back across the bed towards the plants. This phenomenon also occurs if the normally dry furrows are accidentally irrigated.

Sprinkler Irrigation Sprinkler-irrigated fields with poor water quality present a challenge because it is difficult to apply enough water to leach the salts and you cannot effectively utilize row or bed configurations to manage accumulation. Growers should monitor the soil EC and irrigation water salinity. Where adequate irrigation water exists above crop requirements, a leaching fraction (or percent of additional water needed above crop requirements) can be calculated for sprinkler irrigated fields using this equation:

$$\%\,Leaching\ requirement = \frac{ECwater}{2xEC\max}x100$$

In this equation, EC max is the maximum soil EC wanted in the root zone. Apply this leaching fraction to coincide with periods of low soil N and residual pesticide. Again, fall is an optimal time to move salts below the root

The spacing of the drains depends on the level of the land and the permeability of the soil . The drains can be placed in a grid or herring bone fashion, depending upon the level of the water-logged area. In the case of the sugarcane growing areas in Tamil Nadu and Andhra Pradesh the farmers are using mole drains. They are made at a depth of 0.45 to 0.6m and are quite effective. The tractor-mounted trenching machines are very useful in digging the channels for putting the tile drains.

In order to drain the water from the fields the Japanese farmers use vertical pumps having low lifts and high volume. Samples of such pumps are now available in India. The capacity is 136,380 to 181,540 litres per hour and they can be effectively used for draining the soil, if the utility channels are available.

Tractors and Tractor drawn implements Even though we have to lay greater emphasis on manually operated and bullock drawn implements and machinery for the present the slow mechanization of agriculture with power-operated implements has begun to come of age. It is generally considered that the high yield in the western countries is due to the high level of mechanisation of various farm operations.

Mechanisation is possible there due to the shortage of labour and the development of engineering industry. So far these conditions have not existed in India. The labour has been cheap and the industries underdeveloped. Agricultural labour is getting scarce. Almost any implement can be now manufactured cheaply and effectively. The farmers are aware of running the farm as a business and agriculture is no longer a way of subsistence living. The governments at the centre and state are also doing their part by bringing to the farmer the results of research. That mechanisation can succeed has also been shown by the farmers already having adopted diesel engine pumping sets and electrically driven waterlifts. In certain states a large number of farmers are using tractors and tractor-drawn implements. For the following operations mechanized power can be profitably and effectively used.

1. For carrying out heavy jobs such as summer ploughing or reclaiming vast stretches of land. This work cannot be done with manually operated or bullock drawn implements. Many state govts. have their own tractor units which the farmers take on hire.
2. For eradicating weeds like kans, hariali, dub etc which are deep rooted and obnoxious.
3. For clearing jungles and opening new lands
4. For desilting of tanks
5. For big govt. farms, tea or coffee estates, and for using them on collective basis
6. For custom-hiring wherever possible.

All the above types have been done mainly through the govt agencies in the past. The best example is the Central tractor organisation of the Govt of India which functioned from 1947 to some years ago and reclaimed lakhs of acres in central India. Such large-scale operations require heavy crawler type tractors.

A tractor is a 'prime mover'. For attachment to the tractor we must provide the farmers with different types of implements to carry out different types of operations. Because of the higher horse power used the tractors can do more work than the bullocks. The tractor is sturdy so it can work in rough fields and arrangements have to be made to attach different implements to it and for stationary machinery by using the power take off. Thus the tractor is a modification of a motor car to perform agricultural operations including bund-making etc. The tractors can be classified as:

The crawler tractor or track layer. Such tractors are suitable for relatively high draft and low speed work.

They can also be used on uneven ground or soft marshy land where wheels would not work. These tractors are heavier but their weight is evenly distributed on a larger area; hence they do not compact the ground as much as the wheel type tractors do. The horse power ranges from 60 to 500. They are employed for large pieces of barren land to make it fit for cultivation, for preparing barriers or bunds, desilting of tanks, levelling of land for draining away excess water.

Four wheeled tractors of 15-60 horse power. Under this category come most of the agricultural tractors of horse power 30-35. They are fitted with pneumatic tyres. They can be used for ploughing, harrowing, sowing, harvesting, transport and belt work. They can also be used for intercultivation by changing the width between the wheels. By using trailors they can be used for transporting seeds, manure, fertilizer etc . They are also used for belt work like cutting chaff, lifting water and for working a wheat thresher.

Small tractors and power tillers. A small tractor can be defined as a machine having four wheels and 2 axles. A power tiller has 2 wheels or one axle. The horse power ranges from 2 to 15. The Indian farmers are showing a good deal of interest in the small tractors as the price of a good pair of bullocks is alarming and they require expensive maintenance. They use machines in the range of 5 to 10 hp costing ₹ 2000-3000. The main jobs that the machines would be performing are ploughing, harrowing, puddling, transport, pumpimng water and spraying insecticides. Such small machines are also very useful because of the reduced sizes of land holdings. In most states the ceiling on land has been fixed to 10.117 to 12.140 ha and it is not economical for farmers to use bog tractors. The conventional four wheeled tractors can be used only on custom basis or co-operative basis. The four wheeled tractors have the riding system whereas the power tillers are walking type tractors. They are common in japan, Germany, Italy, France etc. They are provided with a variety of attachments which can perform almost any agricultural operation.

Apart from the types of tractors mentioned above there are half trach tractors as well as baby track layers. For wetland paddy cultivation cage wheels have been designed. They are also called skeleton wheels. They can be broad or narrow ranging from 0.5 to 1m in width. They help the tractor to distribute its weight on the wetland and do not allow it to sink. The cage wheels are also useful in puddling.

Drainage has long been an important component of land management in the Coastal Plain and Tidewater regions of North Carolina. On flat, poorly drained soils, intensive drainage is necessary to facilitate seedbed preparation and planting in order to minimize plant stress and subsequent yield reduction resulting from poor soil aeration that accompanies waterlogging.

Nearly half of the cropland currently used in North Carolina requires drainage improvement for efficient production. The drainage intensity required for agricultural production is not the same in all years or all periods of the year. While wetness is the major concern, weather conditions vary such that crops periodically suffer from drought stress that may substantially reduce yields in some years. Intensive drainage systems, necessary to provide trafficability during extreme wet periods often, remove more water than necessary during drier periods, leading to temporary overdrainage (Doty et al., 1986).

Problems with drought on drained soils have resulted in a transition from conventional drainage methods to water table management systems. The latter provide drainage during wet periods, but utilize control structures to manage the water level in the drainage outlet, making it possible to reduce overdrainage. In some cases, the system can be used to provide subirrigation during dry periods. Collectively, these practices are referred to as water table management and involve a combination of management practices including surface drainage, subsurface drainage, controlled drainage, and/ or subirrigation.

Types of Drainage Systems: What They Do and How They Work

Drainage is accomplished by two methods — open-ditch systems designed to provide primarily surface drainage (surface runoff) or underground systems comprised of drain tile or tubing designed to lower the water table by subsurface flow. A surface drainage system typically consist of 3- to 5-foot deep, open ditches installed on 300- to 600-foot intervals. Surface runoff develops when the rate of rainfall exceeds the soil's capacity to absorb water, thereby resulting in surface ponding. Shallow surface drains (hoe drains) are often utilized to effectively convey ponded surface water to ditches. Vegetated field borders and drop inlet pipes are used to stabilize ditch banks and minimize erosion while conveying surface runoff from the surface drains into the ditch.

Subsurface drainage is obtained by buried tile or tubing (4- to 6-inch diameter) that is placed 3 to 5 feet deep and 50 to 200 feet apart. A subsurface system provides drainage when the water table rises above the drain depth and water flows towards and into the drain. The drainage process whereby water infiltrates into the soil and moves within the soil profile is referred to as subsurface drainage, shallow ground water flow, or sometimes interflow.

In practice, it is often difficult to differentiate between surface and subsurface drainage, particularly in eastern North Carolina, because the outflow in drainage ditches or canals is usually a combination of both surface and subsurface flow. The relative proportion of surface and subsurface flow in the total drainage volume depends on many factors. These include rainfall intensity, land surface roughness and slope, vegetation, soil permeability, and ditch or drain tubing spacing and depth.

Open ditches are normally spaced farther apart than buried tubing, which typically causes subsurface flow to be slow, resulting in collection of predominately surface drainage. But in highly permeable soils, open ditches may provide significant subsurface drainage. Such is the case in the Tidewater and Lower Coastal Plain where many fields are underlain by highly permeable sands at shallow depths, typically within 3 to 6 feet of the soil surface. Under such conditions, the drainage in open ditch systems is often predominately subsurface flow even through the ditch system is referred to as a surface drainage system.

Drainage and Water Quality Nitrogen and phosphorus are transported from land-based activities to receiving streams and estuaries by drainage of excess rainfall. The difference in drainage method (surface versus subsurface flow) is important from a water quality standpoint because the characteristics of the two drainage waters differ. Surface drainage systems result in rapid removal of excess water over a relatively short time period. This water flowing over the land surface has relatively high energy sufficient to detach and transport soil particles and constituents attached to them, such as phosphorus, organic nitrogen, and many pesticides (Gilliam et al., 1978; Skaggs and Gilliam, 1981; Deal et al., 1986). Subsurface drainage typically contains very little sediment, but contains high concentrations of soluble constituents such as nitrate-nitrogen (Gilliam et al., 1978; Skaggs and Gilliam, 1981; Skaggs et al., 1982; Evans et al., 1987; Deal et al., 1986). Field research has documented nitrogen losses in drainage water from the edge of agricultural fields to average about 20 lbs. nitrogen acre-1 year-1. Site-to-site and year-to-year variation ranges from none to over 40 lbs. nitrogen acre-1 year-1. Factors causing this variability include land use, type of drainage, drainage intensity, and variability in rainfall, soil, landscape position, fertilization rate, type of crop, and harvested crop yield.

Controlled Drainage Water control structures, such as a flashboard riser, installed in the drainage outlet allow the water in the drainage outlet to be raised or lowered as needed. This water management practice has become known as controlled drainage. When the flashboards are lowered or removed, subsurface drainage occurs more quickly. When flashboards are added to the riser, the subsurface drainage rate is decreased and the height of the water level in the ditches and surrounding fields rises. Managing the field water through the use of controlled drainage allows timely drainage but also maximum storage of water within the field for utilization by the crop.

The transport of nitrogen from drained fields can be minimized by managing the drainage system such that only the minimum drainage water necessary is allowed to exit the field. In numerous field studies, drainage control reduced the annual transport of total nitrogen at the field edge by 9 lbs. acre-1 year-1 or 45 per cent on average.

Nitrogen reductions resulting from controlled drainage result from two processes. First, controlled drainage reduces the volume of drainage water leaving a field from 20–30 per cent on average; however, outflow varies widely depending on soil type, rainfall, type of drainage system and management intensity. During dry years, controlled drainage may totally eliminate outflow.

In wet years, control may have little or no effect on total outflow. Second, controlled drainage provides a higher field water table level which promotes denitrification within the soil profile. In some cases, nitrate-nitrogen concentrations have been 10–20 per cent lower in outflow from controlled systems compared to uncontrolled, free-draining systems. The combined effect of reduced flow and reduced nitrate concentration results in the overall 45 per cent reduction in nitrogen mass transport at the field edge. Controlled drainage has also been documented to reduce phosphorus transport by 0.1 lbs. acre-1 or roughly 35 per cent.

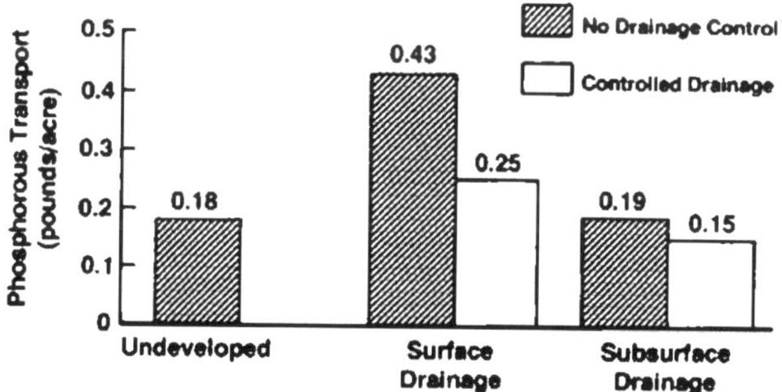

Outflow as measured at the field edge for 12 soils and sites. Values shown are for mineral soils only. Two sites with organic soils were not included.

The successful management of controlled drainage systems rests on two important objectives. The first is achieving optimum production efficiency and maximum nutrient utilization by the crop; the second is attaining maximum water quality benefits. Controlled drainage structures require that the topography be relatively flat. The costs to production and water quality will usually exceed benefits when the land slope exceeds 0.5 per cent. As a consequence, controlled drainage is most practical in the Lower Coastal Plain and Tidewater regions of North Carolina. The predominant cropping sequence in these areas is a two-year rotation of corn, wheat, and soybean. General guidelines for management of controlled drainage under this cropping sequence are given in Table 5.

The general guidelines present an attempt to achieve a balance between production and water quality goals. Many of the management indicators are hidden from view and the response to adjustments is not immediate. Intensive management with long- term monitoring is necessary to develop a site-specific understanding of the system.

Productivity and water goals are compatible during some years, or at least seasonally during most the year. Under some conditions, however, productivity, water quality, or both goals may need to be mutually compromised for the benefit of the other. For example, management throughout the year is necessary to achieve maximum water quality benefit and drainage control to reduce nitrogen transport is most effective during winter and early spring periods. When fields are fallow, there is no significant production cost associated with holding water levels high to achieve the maximum water quality benefit. But productivity of some crops such as winter wheat may be reduced by high winter water table levels.

Table 5. General water table management guidelines to promote water quality and optimum crop yields for a two-year rotation of corn-wheat-soybeans.

Period	Production Activity	Control Setting* (inches)	Comments**
Jan.1-Mar.15	Fallow	12-18	Minimize drainage outflow and encourage denitrification
Mar. 15 - Apr. 5	Tillage, corn seedbed preparation, planting	30-36	Just deep enough to provide trafficability and good conditions for seedbed preparation.
Apr. 15 - May 15	Corn establishment, early growth	24 - 30	Deep enough to promote good root development.
	Nitrogen sidedressing	24-36	Just low enough to allow trafficability
May 15 - Aug. 15	Corn development and maturity	18-24	Temporary adjusting during wet periods
Aug.15 - Oct. 15	Harvesting, tillage; planting of wheat	30-36	Lower enough to provide trafficabilty
Oct. 15 - Mar. 1	Wheat establishment	18-24	Lower during extremely wet periods
Mar.1 - Mar. 15	Sidedressing wheat	30-36	Low enough to provide trafficability
Mar.15 - Jun. 15	Wheat development	18-24	Temporary adjustment during wet periods
Jun. 15 - Jul. 15	Harvesting wheat; tillage, planting of soybeans	30-36	Depends on rainfall
Jul. 15 - Nov. 1	Soybean development and maturity	18-24	Temporary adjustment to allow cultivation
Nov. 1 - Dec. 15	Soybean harvesting	36-42	Low enough to provide trafficability
Dec. 15 - Mar. 15	Fallow	12-18	Minimize drainage outflow and encourage denitrification

*Values shown are the control setting (depth below average surface elevation) and should not be considered the actual water table depth in the field, which will be lower except during drainage periods.

**Most adjustments are related to trafficability and must take into account weather conditions and soil water status at the time:

- In an unusually dry season, control can be 3 to 6 inches higher;
- In an unusually wet season, control should be 3 to 6 inches lower;
- In coarse-textured soils, trafficability can be provided with the water table approximately 6 inches higher.

9

Plant Nutrition

THE ELEMENTS OF COMPLETE PLANT NUTRITION

The following is a brief guideline of the role of essential and beneficial mineral nutrients that are crucial for growth. Eliminate any one of these elements, and plants will display abnormalities of growth, deficiency symptoms, or may not reproduce normally.

MACRONUTRIENTS

- Nitrogen is a major component of proteins, hormones, chlorophyll, vitamins and enzymes essential for plant life. Nitrogen metabolism is a major factor in stem and leaf growth (vegetative growth). Too much can delay flowering and fruiting. Deficiencies can reduce yields, cause yellowing of the leaves and stunt growth.
- Phosphorus is necessary for seed germination, photosynthesis, protein formation and almost all aspects of growth and metabolism in plants. It is essential for flower and fruit formation. Low pH (<4) results in phosphate being chemically locked up in organic soils. Deficiency symptoms are purple stems and leaves; maturity and growth are retarded. Yields of fruit and flowers are poor. Premature drop of fruits and flowers may often occur. Phosphorus must be applied close to the plant's roots in order for the plant to utilise it. Large applications of phosphorus without adequate levels of zinc can cause a zinc deficiency.
- Potassium is necessary for formation of sugars, starches, carbohydrates, protein synthesis and cell division in roots and other parts of the plant. It helps to adjust water balance, improves stem rigidity and cold hardiness, enhances flavour and colour on fruit and vegetable crops, increases the oil content of fruits and is important for leafy crops. Deficiencies result in low yields, mottled, spotted or curled leaves, scorched or burned look to leaves..

- Sulphur is a structural component of amino acids, proteins, vitamins and enzymes and is essential to produce chlorophyll. It imparts flavour to many vegetables. Deficiencies show as light green leaves. Sulphur is readily lost by leaching from soils and should be applied with a nutrient formula. Some water supplies may contain Sulphur.
- Magnesium is a critical structural component of the chlorophyll molecule and is necessary for functioning of plant enzymes to produce carbohydrates, sugars and fats. It is used for fruit and nut formation and essential for germination of seeds. Deficient plants appear chlorotic, show yellowing between veins of older leaves; leaves may droop. Magnesium is leached by watering and must be supplied when feeding. It can be applied as a foliar spray to correct deficiencies.
- Calcium activates enzymes, is a structural component of cell walls, influences water movement in cells and is necessary for cell growth and division. Some plants must have calcium to take up nitrogen and other minerals. Calcium is easily leached. Calcium, once deposited in plant tissue, is immobile (non–translocatable) so there must be a constant supply for growth. Deficiency causes stunting of new growth in stems, flowers and roots. Symptoms range from distorted new growth to black spots on leaves and fruit. Yellow leaf margins may also appear.
- Carbon forms the backbone of many plants biomolecules, including starches and cellulose. Carbon is fixed through photosynthesis from the carbon dioxide in the air and is a part of the carbohydrates that store energy in the plant.
- Hydrogen also is necessary for building sugars and building the plant. It is obtained almost entirely from water. Hydrogen ions are imperative for a proton gradient to help drive the electron transport chain in photosynthesis and for respiration.
- Oxygen is necessary for cellular respiration. Cellular respiration is the process of generating energy–rich adenosine triphosphate (ATP) via the consumption of sugars made in photosynthesis. Plants produce oxygen gas during photosynthesis to produce glucose but then require oxygen to undergo aerobic cellular respiration and break down this glucose and produce ATP.

Micronutrients:

- Iron is necessary for many enzyme functions and as a catalyst for the synthesis of chlorophyll. It is essential for the young growing parts of plants. Deficiencies are pale leaf colour of young leaves followed by yellowing of leaves and large veins. Iron is lost by leaching and is held in the lower portions of the soil structure. Under conditions of high pH (alkaline) iron is rendered unavailable to plants. When soils are alkaline, iron may be abundant but unavailable. Applications of an acid nutrient formula containing iron chelates, held in soluble form, should correct the problem.

- Manganese is involved in enzyme activity for photosynthesis, respiration, and nitrogen metabolism. Deficiency in young leaves may show a network of green veins on a light green background similar to an iron deficiency. In the advanced stages the light green parts become white, and leaves are shed. Brownish, black, or grayish spots may appear next to the veins. In neutral or alkaline soils plants often show deficiency symptoms. In highly acid soils, manganese may be available to the extent that it results in toxicity.
- Boron is necessary for cell wall formation, membrane integrity, calcium uptake and may aid in the translocation of sugars, oron affects at least 16 functions in plants. These functions include flowering, pollen germination, fruiting, cell division, water relationships and the movement of hormones. Boron must be available throughout the life of the plant. It is not translocated and is easily leached from soils. Deficiencies kill terminal buds leaving a rosette effect on the plant. Leaves are thick, curled and brittle. Fruits, tubers and roots are discoloured, cracked and flecked with brown spots.
- Zinc is a component of enzymes or a functional cofactor of a large number of enzymes including auxins (plant growth hormones). It is essential to carbohydrate metabolism, protein synthesis and internodal elongation (stem growth). Deficient plants have mottled leaves with irregular chlorotic areas. Zinc deficiency leads to iron deficiency causing similar symptoms. Deficiency occurs on eroded soils and is least available at a pH range of 5.5—7.0. Lowering the pH can render zinc more available to the point of toxicity.
- Copper is concentrated in roots of plants and plays a part in nitrogen metabolism. It is a component of several enzymes and may be part of the enzyme systems that use carbohydrates and proteins. Deficiencies cause die back of the shoot tips, and terminal leaves develop brown spots. Copper is bound tightly in organic matter and may be deficient in highly organic soils. It is not readily lost from soil but may often be unavailable. Too much copper can cause toxicity.
- Molybdenum is a structural component of the enzyme that reduces nitrates to ammonia. Without it, the synthesis of proteins is blocked and plant growth ceases. Root nodule (nitrogen fixing) bacteria also require it. Seeds may not form completely, and nitrogen deficiency may occur if plants are lacking molybdenum. Deficiency signs are pale green leaves with rolled or cupped margins.

- Chlorine is involved in osmosis, the ionic balance necessary for plants to take up mineral elements and in photosynthesis. Deficiency symptoms include wilting, stubby roots, chlorosis (yellowing) and bronsing. Odours in some plants may be decreased. Chloride, the ionic form of chlorine used by plants, is usually found in soluble forms and is lost by leaching. Some plants may show signs of toxicity if levels are too high.
- Nickel has just recently won the status as an essential trace element for plants according to the Agricultural Research Service Plant, Soil and Nutrition Laboratory in Ithaca, NY. It is required for the enzyme urease to break down urea to liberate the nitrogen into a usable form for plants. Nickel is required for iron absorption. Seeds need nickel in order to germinate. Plants grown without additional nickel will gradually reach a deficient level at about the time they mature and begin reproductive growth. If nickel is deficient plants may fail to produce viable seeds.
- Sodium is involved in osmotic (water movement) and ionic balance in plants.
- Cobalt is required for nitrogen fixation in legumes and in root nodules of non-legumes. The demand for cobalt is much higher for nitrogen fixation than for ammonium nutrition. Deficient levels could result in nitrogen deficiency symptoms.
- Silicon is found as a component of cell walls. Plants with supplies of soluble silicon produce stronger, tougher cell walls making them a mechanical barrier to piercing and sucking insects. This significantly enhances plant heat and drought tolerance. Foliar sprays of silicon have also shown benefits reducing populations of aphids on field crops. Tests have also found that silicon can be deposited by the plants at the site of infection by fungus to combat the penetration of the cell walls by the attacking fungus. Improved leaf erectness, stem strength and prevention or depression of iron and manganese toxicity have all been noted as effects from silicon. Silicon has not been determined essential for all plants but may be beneficial for many.

MINOR ELEMENTS IN PLANT NUTRITION

Copper: Copper has long been known as an element toxic to plants even when present in what may be considered high dilution. One part in 10 million of water is toxic to fresh-water algae and copper sulphate is commonly used in the elimination of water weeds from tanks, etc. This element is receiving more and more attention as a minor element. Its increasing importance arises from the fact that copper deficiency causes pathological symptoms in animals as well as in plants. In the higher plants a modicum of copper is essential, and many soils, often those of a peaty character, fail to provide the very small amount necessary.

The addition of to of one part per million in water culture, or about 1 per cent to 3 Ib. of copper per acre, seems to eliminate all trouble in this respect. The element is often supplied in the form of the sulphate or pyrites: the amount of either of these to be sown depends on their copper content. The mechanism through which the copper acts is not known. That plants deficient in copper are often lacking in chlorophyll is not without importance, for copper deficiency in the diet of infants is often the cause of some of the anaemias that they suffer from.

The red pigment of blood and the green pigment of the plant are related in very many respects. Where copper is in too great supply it has been shown that iron acts as a "depoisoner." Deficiency of zinc, too, causes pathological symptoms in the plant. It is thought that the curative action of some fungicidal zinc sprays is more connected with the action of the metal on the physiology of the host that its direct effect on the disease organism.

Cobalt: Soils deficient in cobalt produce herbage of normal appearance, but the animals fed on the produce pine and die. The addition of about 2 Ib. cobalt chloride per acre eliminates the trouble.

Boron: Of all the minor elements probably none has received so much attention as boron. It is deficient in quantity in many soils all over the world. Symptoms of disfunction following from apparent boron deficiency seem to be connected with calcium supply, and show fundamentally in reduction or stoppage of activity in the meristematic regions. Further, plants showing symptoms of boron deficiency are found usually in calcareous soils or those which have been heavily limed. Boron does not seem to be necessary to animals, so that the chief interest in this minor nutrient resides in the pathological condition it may set up in the plant.

Manganese: Manganese deficiency symptoms, too, are most commonly found in plants growing on soils well supplied with lime. Except at high dilution manganese is toxic to plants, yet when presented in too low a concentration metabolic upsets result. These are definitely connected with lowered efficiency of the respiratory function.

Indeed manganese seems to be necessary for the whole oxidation/ reduction mechanism. Plants offered a nitrogen supply in the form of ammonia compounds are benefited by extremely small additions of manganese to the nutrient medium.

This is especially so when the oxygen supply is limited. Additions of manganese seem to benefit the legumes more than other cultivated plants. A sufficiency of manganese in the food is necessary for animals.

Molybdenum: Molybdenum was fairly early shown to be necessary for the nutrition of many fungi, but it is now thought that higher plants use it. Certain soils, typically those lying on the Lower Lias formation in southeast England, contain about 20 to 100 parts per million of this element, and the herbage produced on them may contain about 2 to 3 parts per million. When grazing on such a pasture ruminant animals scour, pine, and eventually die. Such pastures are known as "teart." Clovers seem to absorb more molybdenum than grasses, hence, increase of these valuable plants intensifies the bad effect.

It would seem that the "availability" of molybdenum is controlled by the acidity of the soil. In neutral or alkaline soils the availability is high; in acid conditions low. Addition of lime to the soil does not render teart pastures healthy, but rather the reverse.

Also, the clovers being encouraged by the lime become more prolific, and because of their high molybdenum content they raise the content of the herbage as a whole. There is some evidence that copper may be antagonistic to molybdenum and be useful in treating affected animals or in improving teart pastures. Other elements affect the plant and through the plant, animals. Enough has been said to elucidate the principles involved, and for fuller details of each case reference should be made to the very voluminous literature which has already grown up.

COMPOUNDS OF ELEMENTS

Selenium: It will be convenient to discuss -one of the non-essential elements here as it is intimately related to sulphur. This is selenium, which is of special interest because it is often absorbed by plants in a quantity sufficient to render them toxic to animals while themselves remaining unaffected. The element is comparatively rare, but traces of its compounds in the soil result in herbage, grain, etc., producing all the symptoms of the dreaded alkali disease when fed to animals.

Man, too, is subject to selenium poisoning. There is no doubt that plants "heap up" selenium, and that the accumulation is due in large part to the metal replacing the sulphur in certain of the amino-acids of the protein molecule. The process of "heaping up" need not go very far for the plant to become toxic to animals. Rats die when fed on perfectly normal-appearing plants containing 2 or 3 parts per million of selenium. On the other hand, "heaping-up" must go very far before the plants show symptoms. Fifteen parts per million in the plant produces chlorosis, and with greater dosage the leaves and roots develop a pink coloration.

Fortunately selenium and sulphur are antagonistic. Addition of sulphur to the soil substantially reduces the amount of selenium subsequently found in the tissues of the plant. By applications of gypsum, selenium-containing soil can be made to produce crops which are non-toxic to animals.

Selenium-tolerant Plants: Some plants not only tolerate this element so toxic to animals but actually require it for successful growth. *Astragalus* *racemosus* and *A. pattersonia* are reported as thriving in a medium containing 27 parts of selenium per million. This is about ten times the amount tolerated by wheat or buckwheat.

Furthermore, in these plants addition of sulphur to the soil, while resulting in considerable reduction of the selenium content, does not bring it down to a level likely to be tolerated by grazing animals. Plants which occur naturally on seleniferous soils and nowhere else are now becoming known. It is expected that delimitation of seleniferous areas by mapping the occurrence and distribution of these "indicator" plants will be comparatively simple, and so eliminate much chemical work on soil analysis.

Selenium as an Insecticide: An interesting development of these discoveries regarding selenium is that crops can be protected from biting or sucking insects if selenium is added to the soil in which the plants grow. Insects die after feeding on the seleniferous juice. This can be made use of without detriment to man or his animals only when the harvested produce is not consumed as food, *e.g.,* timber and fibres, or where it is food but not protein, *e.g.,* sugar. A slight exception to this is seen in cacao where the very young plant is subject to attacks from insect pests.

The juvenile plantlets are supplied for a short period with selenium, after which they are transplanted to fresh untreated land. By the time the plants are adult and bearing a crop the small amount of this element which protected the small plant is dispersed through a large plant, and the small percentage then present is not toxic in the product.

Phosphorus: Phosphorus is of fundamental importance in the economy of the plant, involved as it is in protein formation, as well as in respiration and other functions. The utilisation of nitrogen by the plant is interfered with if phosphorus is in deficient supply. This may be due directly to insufficiency of phosphate for the formation of protein. There is evidence, however, that the action lies deeper, for when phosphates are deficient the enzyme responsible for the initial steps in nitrate utilisation is either absent or inoperative. Lack of phosphate seems, then, to "bottleneck "the whole of the synthesis in the plant.

Mono- and di-saccharides "bank-up" amino and amide nitrogen accumulate, and eventually cell-division slows down and ceases. The slowing down of cell-division may be a reflex of deficiency of necessary phosphates for protoplasm formation, but it is also known that When phosphorus supply is inadequate the fibres which "pull" the split chromosomes apart in mitosis are not formed.

A concomitant of this whole imbalance of cell function is cessation of all phenomena requiring active cell-division, *e.g.,* growth of the axis and branching. This is particularly apparent in cereals where the result is that tillering is depressed. The' converse of these effects is seen when phosphate is supplied liberally, for there is then more active branching of the root and shoot, while the plant tends to flower and seed not only earlier but more profusely.

Calcium: The predominant role of calcium in agriculture is not wholly botanical, but arises from the use of lime as a soil ameliorant and as a reducer of soil acidity.

The element, however, is taken into the plant in quite substantial quantities, probably as the metallic ion. In the tissues it neutralises acids arising from many metabolic activities, and crystals of the calcium salts of oxalic and other organic acids commonly appear in cells. One of the principal roles of calcium in the plant seems to be involved in some fine balance with boron, and possibly other elements.

What the relationship is still remains obscure. Certainly deficiencies of both boron and calcium are very similar in the symptoms they produce. This is especially so in the failure of normal development in meristematic regions. Calcium is found in great quantities in leaves, and does not migrate much from an old organ to a young one as the plant develops. Calcium is an essential ingredient of the structure of the cell wall, or at least of the middle lamella.

Nitrogen: Let us now turn to the special characteristics of each element in connection with intake. It was believed at one time that the plant used nitrogen only as the nitrate ion NO_3, and that ammonia and other forms had to be converted to this radicle in the soil before entry or utilisation was possible. It is now known that in the field ammonia is used satisfactorily by some plants.

Whether any particular kind of plant can make use of ammonia or not would seem to depend on whether it can "step up" this form of nitrogen quickly to higher combinations such as an amide. Apparently ammonia cannot be stored as such in the cells. The rate of intake of ammonia seems to be faster than the intake of nitrate—at least in such plants as the cotton seedling, oats, and buckwheat. Whether ammonia or nitrate is the more suiform seems to be governed by at least two factors, the nature of the plant and the reaction of the soil.

An acid soil favours the use of nitrate. The successful use of nitrite (NO_2) by plants has been reported, but is of little agricultural interest. The chief use of nitrogen in the plant is in the formation of compounds such as amino-acids, which enter into the molecular structure of the proteins. Since, proteins are the basis of protoplasm, reduced nitrogen supply soon shows itself in reduction of protoplasm formation.

This puts a restriction on practically all functions, and particularly on new cell formation, so that growth slows and may cease.

Sulphur: Sulphur is used by the plant only when offered in the form of a sulphate (SO_4); most other forms are toxic or not usable. The sulphur enters into certain amino-acids which are essential in the formation of many proteins. Many plants can maintain apparently normal growth and appearance while suffering from partial lack of sulphur. The tissues are, however, deficient in the element.

A peculiar case of sulphur deficiency occurs in some grassland areas only in spring when growth of the grass is very rapid. The sulphur intake at this time is not sufficient to supply what is necessary for normal protein building, but the plant goes on growing, using such sulphur as it has very economically, or improvising in some other way. A mass of herbage deficient in sulphur is produced. Later in the year, when growth slows down, the sulphur supply is sufficient, so that the maladjustment corrects itself, and normal protein formation follows.

Break in Wool: As wool protein contains much sulphur, sheep grazing in such an area during the spring develop a peculiar characteristic of the wool. When a lock of the wool is sharply pulled, each fibre snaps at the same point, and the broken tuft shows a clean break instead of the irregular break obtained with normal wool. When wool fibres of this type are seen under the microscope an identical part of each fibre is seen to be of reduced diameter. This is the weak point, and is where the break takes place.

Working from known development rates for wool fibre, it appears that the weak part is produced while the sheep is feeding on the spring flush of grass, which is deficient in sulphur. The condition is remedied by feeding during the deficiency period a "lick" of hydrolised wool, which supplies organic sulphur.

Potassium: Potassium in plant nutrition offers something of a paradox in that commonly there is a considerable quantity of potash in the soil with only a small proportion soluble, while in the plant practically the whole of the potassium remains soluble in the cell sap. When a plant tissue is mashed up the potassium may be easily leached out. Despite its appearance in this simple form potassium has fundamental roles to perform. Potassium is concerned with the synthesis of carbohydrates, and probably with their translocation.

This connection between potassium and carbohydrates is clearly seen in crops which normally accumulate large reserves of starch or sugar. These crops when grown in a deficiency of sunlight produce less reserve food (lower crop yield), but when generously supplied with potash the production is greatly increased.

It would appear that potassium can "replace" sunlight. Potassium is moved out of adult tissues to areas of meristematic activity. Potash confers on plants some degree of resistance to various diseases. The clovers and their allies, though not "carbohydrate" plants, gain greater benefit from increased potash supplies than do grasses, and hence, in pastures with a grass/clover sward addition of potash tends to strengthen clovers rather than the grasses.

Magnesium: Magnesium is an essential ingredient for the formation of chlorophyll, and indeed is the only metallic constituent of this all-important green pigment of plants. As the plant passes to final maturity and the leaves change colour, magnesium is largely transferred from the vegetative body to the seed.

In the plant another of its functions is connected with its use along with phosphate in the formation of those special proteins which form the cell nucleus (the nucleo-proteins). Magnesium, too, appears to be concerned in oil metabolism. The amount of magnesium required by the plant is not large and, when too great quantity is available in the soil functional disturbances develop in the plant.

Iron: Iron is not an essential element in the sense that the term has so far been used, for only a little is required. Neither is it regarded as a minor element, but occupies an intermediate position in the classification. Iron seems to take no structural part in the plant, yet its presence is essential for the formation of chlorophyll.

Though necessary for the formation of chlorophyll, it does not enter into the molecule of that substance. Iron is not mobile inside the plant. This is easily shown by first growing plants for a time in a solution supplying iron when the shoot is fully green.

If the experimental plants are then transferred to a culture medium deficient in iron, the new leaves produced in further growth are white due to a lack of iron in them inhibiting formation of chlorophyll. Total absence of iron from soils is not so common as total or nearly total non-availability. Thus it is that many soils rich in iron produce plants showing all the symptoms of iron deficiency.

The Calcium/Phosphorus Ratio: The importance in the diet of animals of adequate quantities of calcium and phosphorus individually, and of the ratio of these to each other, needs no stressing here. The data offered in the general account of salt intake should be again examined from this aspect of the quality of food.

CHEMICAL CHANGES DURING PLANT GROWTH

The simple compounds which the plant absorbs from the atmosphere and soil are elaborated within its system, and converted into the various complex substances of which its tissues are composed, by a series of changes, the details of which are still in some respects imperfectly known, although their general nature is sufficiently well understood.

They may be best rendered intelligible by reference, in the first instance, to the changes occurring during germination, when the young plant is nourished by a supply of food stored up in the seed, in sufficient quantity to maintain its existence until the organs by which it is afterwards to draw its nutriment from the air and soil are sufficiently developed to serve that purpose.

CHANGES OCCURRING DURING GERMINATION

When a seed is placed in the soil under favourable circumstances, it becomes the seat of an important and remarkable series of chemical changes, which result in the production of the young plant.

Experiment and observation have shown that heat, moisture, and air, are necessary to the production of these changes, and though probably not absolutely essential, the absence of light is favourable in the early stages. The temperature required for germination varies greatly in different seeds, some germinating readily at a few degrees above the freezing point, and others requiring a tolerably high temperature.

The rapidity with which it takes place appears to increase with the temperature; but this is true only within very narrow limits, for beyond a certain point heat is injurious, and when it exceeds 120° or 130° Fahrenheit, entirely prevents the process. The presence of oxygen is also essential, for it has been shown that if seeds are placed in a soil exposed to an atmosphere deprived of that element, or if they be buried so deep that the air does not reach them, they may lie without change for an unlimited period; but so soon as they are exposed to the air, germination immediately commences.

Illustrations of this fact are frequently observed where earth from a considerable depth has been thrown up to the surface, when it often becomes covered with plants not usually seen in the neighbourhood, which have sprung from buried seeds.

When all the necessary conditions for germination are fulfilled, the seed absorbs moisture, swells up, and sends out a shoot which rises to the surface, and a radicle which descends—the one destined to develop the leaves, the other the roots, by which the plant is afterwards to derive its nutriment from the air and the soil. But until these organs are properly developed, the plant is dependent on the matters contained in the seed itself.

These substances are mostly insoluble, but are brought into solution by the atmospheric oxygen acting upon the gluten, and converting it into a soluble substance called diastase, which in its turn reacts upon the starch, converting it first into dextrine, and then into cellulose, and the latter is finally deposited in the form of organised cells, and produces the first little shoot of the plant.

At the first moment of germination, the oxygen absorbed appears simply to oxidise the constituents of the seed, but this condition exists only for a very limited period, and is soon followed by the evolution of carbonic acid, water being at the same time formed from the organic constituents of the seed, which gradually diminishes in weight. The amount of this diminution is different with different plants, but always considerable.

Boussingault found that the loss of dry substance in the pea amounted in 26 days to 52 per cent, and in wheat to 57 per cent in 51 days. Against this, of course, is to be put the weight of the young plant produced; but this is never sufficient to counterbalance the diminished weight of the seed, for Saussure found that a horse bean and the plant produced from it weighed, after 16 days, less by 29 per cent than the seed before germination.

The same phenomenon is observed in the process of malting, which is in fact the artificial germination of barley, the malt produced always weighing considerably less than the grain from which it was obtained. It was believed by Saussure, and the older investigators, that the carbonic acid evolved was entirely produced from starch and sugar; and as these substances may be viewed as compounds of carbon and water, the change was very simply explained by supposing that the carbon was oxidised and converted into carbonic acid and its water eliminated.

But this hypothesis is incapable of explaining all the phenomena observed; for woody fibre, which is one of the chief constituents of the young plant, contains more carbon than the starch and sugar from which it must have been produced, and we are, therefore, forced to admit that the action must be more complicated.

There is every reason to believe that the nitrogenous constituents of the seed are most abundantly oxidised, for they are remarkably prone to change; but the action of the air is not confined to them, and it appears most probable that all the substances take part in the decomposition, and the process of germination may, in some respects, be compared to decay or putrefaction, which, like it, is attended by the absorption of oxygen and evolution of carbonic acid; but while in the latter case the residual substances remain in a useless state, in the former they at once become part of a new organism.

Changes Occurring during the After-growth of the Plant: When the plant has developed its roots and leaves, and exhausted the store of materials laid up for it in the seed, it begins to derive its subsistence from the surrounding air, and to absorb carbonic acid, water, ammonia, and nitric acid, and to decompose and convert them into the different constituents of its tissues.

These changes take place slowly at first, and more rapidly as the organs fitted for the elaboration of its food are developed. The roots and the leaves are equally active in performing this duty, the former absorbing the mineral matters along with the carbonic acid, ammonia, nitric acid, and moisture in the soil, or the manure added to it; the latter gathering the gaseous substances existing in the air. Each of these undergoes a series of changes claiming our consideration.

Decomposition of Carbonic Acid: Carbonic acid, which appears to be absorbed with equal readiness by the roots, leaves, and stems, undergoes immediate decomposition, its carbon being retained, and its oxygen, in whole or in part, evolved into the air. This decomposition occurs only under the action of the sun's rays, and has been found to be proportionate to the amount of light to which the plant is exposed.

It takes place only in the green parts of plants, for though the roots absorb carbonic acid, they cannot decompose it, or evolve oxygen; and the coloured parts, the flowers, fruits, etc., have an entirely opposite effect, absorbing oxygen and giving off carbonic acid.

The absorption of carbonic acid and escape of oxygen has been proved by numerous direct experiments by Saussure and others, in which both atmospheric air and artificial mixtures containing an increased quantity of carbonic acid have been employed.

Saussure allowed seven plants of periwinkle (Vinca minor) to vegetate in an atmosphere containing 7·5 per cent of carbonic acid for six days, during each of which the apparatus was exposed for six hours to the sun's rays.

In this experiment the whole of the carbonic acid, amounting to 431 volumes, was absorbed, but only 292 volumes of oxygen were given off. Had the carbonic acid been entirely decomposed, and all its oxygen eliminated, its volume would have been equal to that of the acid, or 431, so that in this instance 139 volumes of the oxygen of the carbonic acid have been retained to form part of the tissues of the plant.

On the other hand, the nitrogen is found to be increased after the experiment. It might be supposed that the nitrogen evolved had been derived from the decomposition of the nitrogenous constituents of the plant, but this cannot be the true explanation, because in this particular case it greatly exceeded the whole nitrogen contained in the plants experimented on. Its source is not well understood, but Boussingault supposes it to have existed in the interstices of the plant, and to have escaped during the course of the experiment. Saussure found that the oak, the horse-chesnut, and other plants, absorb oxygen and give off carbonic acid in less volumes than the oxygen, while the house-leek and the cactus absorb oxygen without evolving carbonic acid.

The absorption and decomposition of carbonic acid takes place only during the day, and matters are entirely reversed during the night, when oxygen is absorbed and carbonic acid eliminated from all parts of the plants. Although the action occurring during the night is the reverse of that which takes place during the day, it is in no degree to be attributed to a re-oxidation of the carbon which had been deposited in the tissues of the plant.

It appears, on the contrary, to be a purely mechanical, and not a chemical process. During the night the sap continues to circulate through the vessels of the plant, and moisture, carrying with it carbonic acid in solution, is absorbed by the roots; but when it reaches the leaves, where the sun's light would have caused its decomposition during the day, it is again exhaled unchanged. The oxygen absorbed during the night must, however, take part in some chemical processes, for if it were merely mechanical, the absorption would not be confined to that gas alone, but would be participated in by the other constituents of the air. Moreover, the amount of absorption varies greatly in different plants—being scarcely appreciable in some, and very abundant in others.

Plants containing volatile oils, which are readily converted into resins by the action of oxygen, or those containing tannin or other readily oxidisable substances, take up the largest quantity. This is remarkably illustrated by an experiment in which the leaves of the Agave Americana, after twenty-four hours' exposure in the dark, were found to have absorbed only 0·3 of their volume of oxygen, while those of the fir, in which volatile oil is abundant, had taken up twice, and those of the oak, containing tannin, eighteen times as much oxygen.

In the flowers, both by day and night, there is a constant absorption of oxygen, and evolution of carbonic acid. In fact, an active oxidation is going on, attended by the evolution of heat, which, in the Arum maculatum and some other plants, is so great as to raise the temperature of the flower 10° or 12° above that of the surrounding air.

Decomposition of Water in the Plant: In addition to the function which water performs in the plant, as the solvent of the different substances which form its nutriment, and hence, as the medium through which they pass into its organs, it serves also as a direct food, undergoing decomposition, and yielding hydrogen to the organic substances.

Its constituents, along with those of the carbonic acid absorbed, undergo a variety of transformations, and form the principal part of the non-nitrogenous constituents.

It has been already observed that starch, sugar, and the other allied substances, may be considered as compounds of carbon with water; and they might be supposed to owe their origin to the carbonic acid losing the whole of its oxygen, and direct combination then ensuing between the residual carbon and a certain proportion of water; but this would imply that the latter substance undergoes no decomposition, and though undoubtedly the simplest view of the case, it is by no means the most probable.

It is much more likely that the carbonic acid is only partially decomposed, half its oxygen being separated, and replaced by hydrogen, produced by the decomposition of a certain quantity of water into its elements. It must not be supposed that we are in a condition to assert that sugar is really produced in the manner here shown, the illustration being given merely for the purpose of pointing out how it may be supposed to occur, and on a similar principle it is possible to explain the formation of most other vegetable compounds; and this subject has been very fully discussed by the late Dr. Gregory, in his "Handbook of Organic Chemistry."

That water must be decomposed, is evident from the fact, established by analysis, that the hydrogen of the plant generally exceeds the quantity required to form water with its oxygen, so that this excess at least must be produced by the decomposition of water. The hydrogen of the volatile oils, many of which contain no oxygen, and that of the fats, which contain only a small quantity, must manifestly be obtained in a similar manner.

Decomposition of Ammonia: The nitrogenous or albuminous compounds of vegetables must necessarily obtain their nitrogen from the decomposition either of ammonia or nitric acid, experiment having distinctly shown that they are incapable of absorbing it in the free state from the atmosphere.

It has been clearly ascertained that the albuminous substances do not contain ammonia, and it is hence, apparent that a complete decomposition of that substance must take place in the plant. No doubt carbonic acid and water take part with it in these changes, which must be of a very complex character, and in the present state of our knowledge it seems hopeless to attempt any explanation of them.

Decomposition of Nitric Acid: Chemists are not entirely at one as to whether nitric acid is directly absorbed by the plant, or is first converted into ammonia. But there are certain facts connected with the chemistry of the soil, to be afterwards referred to, which seem to us to leave no doubt that it may be directly absorbed; and in that case it must be decomposed, its oxygen being eliminated, and the nitrogen taking part with carbon and hydrogen in the formation of the organic compounds.

It must be clearly understood that while such changes as those described manifestly must take place, the explanations of them which have been attempted by various chemists are not to be accepted as determinately established *facts*; they are at present no more than hypothetical views which have been expressed chiefly with the intention of presenting some definite idea to the mind, and are unsupported by absolute proof; they are only inferences drawn from the general bearings of known facts, and not facts themselves. Although, therefore, they are to be received with caution, they have advantages in so far as they present the matter to us in a somewhat more tangible form than the vague general statements which are all that could otherwise be made.

NITROGENOUS OR ALBUMINOUS CONSTITUENTS OF PLANTS AND ANIMALS

The nitrogenous constituents of plants and animals are so closely allied, both in properties and composition, that they may be most advantageously considered together. ***Albumen:*** Vegetable albumen is found dissolved in the juices of most plants, and is abundant in that of the potato, the turnip, and wheat. In these juices it exists in a soluble state, but when its solution is heated to about 150°, it coagulates into a flocky insoluble substance. It is also thrown down by acids and alcohol.

Coagulated albumen is soluble in alkalies and in nitric acid. Animal albumen exists in the white of eggs, the serum of blood, and the juice of flesh; and from all these sources is scarcely distinguishable in its properties from vegetable albumen.

It is a substance of very complicated composition, and chemists are not agreed as to the formula by which its constitution is to be expressed, a difficulty which occurs also with most of the other nitrogenous compounds. Closely allied to vegetable albumen is the substance known by the name of glutin, which is obtained by boiling the gluten of wheat with alcohol. It appears to be a sort of coagulated albumen, with which its composition completely agrees.

Vegetable Fibrine: If a quantity of wheat flour be tied up in a piece of cloth, and kneaded for some time under water, the starch it contains is gradually washed out, and there remains a quantity of a glutinous substance called gluten. When this is boiled with alcohol, the glutin above referred to is extracted, and vegetable fibrine is left. It dissolves in dilute potash, and on the addition of acetic acid is deposited in a pure state. Treated with hydrochloric acid, diluted with ten times its weight of water, it swells up into a jelly-like mass.

When boiled or preserved for a long time under water, it cannot be distinguished from coagulated albumen.

Caseine: Vegetable caseine exists abundantly in most plants, especially in the seeds, and remains in the juice after albumen has been precipitated by heat, from which it may be separated in flocks by the addition of an acid. It has been obtained for chemical examination, principally from peas and beans, and from the almond and oats.

When prepared from the pea it has been called legumine, from almonds emulsine, and from oats avenine; but they are all three identical in their properties, although formerly believed to be different, and distinguished by these names.

Vegetable caseine is best obtained by treating peas or beans with hot water, and straining the fluid. On standing, the starch held in suspension is deposited, and the caseine is retained in solution in the alkaline fluid; by the addition of an acid it is precipitated as a thick curd. Caseine is insoluble in water, but dissolves readily in alkalies; its solution is not coagulated by heat, but, on evaporation, becomes covered with a thin pellicle, which is renewed as often as it is removed.

Animal Caseine: Animal Caseine is the principal constituent of milk, and is obtained by the cautious addition of an acid to skimmed milk, by which it is precipitated as a thick white curd. It is also obtained by the use of rennet, and the process of curding milk is simply the coagulation of its caseine. It is soluble in alkalies, and precipitated from its solution by acids, and in all other respects agrees with vegetable caseine.

The composition of animal caseine has been well ascertained, but considerable doubt still exists as to that of vegetable caseine, owing to the difficulty of obtaining it absolutely pure.

Other results differ considerably from these, and some observers have even obtained as much as eighteen per cent of nitrogen and fifty-three of carbon. The composition of animal caseine differs from this principally in the amount of carbon.

The most cursory examination of these analytical numbers is sufficient to show that a very close relation subsists between the different substances just described. Indeed, with the exception of vegetable caseine, they may be said all to present the same composition; and, as already mentioned, there are analyses of it which would class it completely with the others. While, however, the quantities of carbon, hydrogen, nitrogen, and oxygen are the same, differences exist in the sulphur and phosphorus they contain, and which, though very small in quantity, are indubitably essential to them.

Much importance has been attributed to these constituents by various chemists, and especially by Mulder, who has endeavoured to make out that all the albuminous substances are compounds of a substance to which he has given the name of proteine, with different quantities of sulphur and phosphorus.

The composition of proteine, according to his newest experiments, is— Carbon 54·0 Hydrogen 7·1 Nitrogen 16·0 Oxygen 21·4 Sulphur 1·5 - 100·0 and is exactly the same from whatever albuminous compound it is obtained. Although the importance of proteine is probably not so great as Mulder supposed, it affords an important illustration of the close similarity of the different substances from which it is obtained, the more especially as there is every reason to believe that the different albuminous compounds are capable of changing into one another, just as starch and sugar are mutually convertible; and the possibility of this change throws much light on many of the phenomena of nutrition in plants and animals.

Indeed, it would seem probable that these compounds are formed from their elements by plants only, and are merely assimilated by animals to produce the nitrogenous constituents they contain. Diastase is the name applied to a substance existing in malt, and obtained by macerating that substance with cold water, and adding a quantity of alcohol to the fluid, when the diastase is immediately precipitated in white flocks. It is produced during the malting process, and is not found in the unmalted barley. Its chemical composition is unknown, but it is nitrogenous, and is believed to be produced by the decomposition of gluten. If a very small quantity of diastase be mixed with starch suspended in hot water, the starch is found gradually to dissolve, and to pass first into the state of dextrine, then into that of sugar.

The change thus effected takes place also in a precisely similar manner in the plant, diastase being produced during the process of germination of all seeds and tubers, for the purpose of effecting this change, and to fulfil other functions less understood, but no doubt equally important. Diastase is found in the seeds only during the period when the starch they contain is passing into sugar; as soon as that change has taken place, its function is ended, and it disappears.

PLANT ESSENTIAL NUTRIENTS

Sixteen chemical elements are recognised as being essential for the growth of all plants. Five others, silicon, sodium, cobalt, vanadium, and nickel, have been recognised as necessary for the growth of some plant species. Although certain essential elements can exist in nature in a number of ionic forms, plants can use only specific ones.

NITROGEN

Other than carbon, hydrogen, and oxygen, nitrogen is the nutrient required by plants in the greatest quantity. The nitrogen concentration of plants ranges from about 0.5 to 5 per cent on a dry weight basis. Since, most plants have a rather high nitrogen requirement and most soils can't supply sufficient nitrogen to meet this demand, nitrogen normally must be supplemented from organic or inorganic fertilizer sources.

The ultimate source of all nitrogen in soils is the atmosphere, which is approximately 78 per cent N_2.

Although this quantity represents an almost unlimited supply, nitrogen as N_2 is not directly available for uptake by most plants. Recall that legumes in symbiosis with particular species of the bacterial genus, ***Rhizobium***, transform gaseous N_2 to plant available form, with capacities to fix N_2 ranging from about 40 to greater than 300 lbs N/acre/year. Non-symbiotic fixation of N_2 by free living soil microorganisms also occurs but quantities are usually less than 10 lbs N/acre/year.

Lightning discharge in thunderstorms can oxidise atmospheric N_2 to plant available nitrate, but quantities are generally less than 20 lbs N/acre/year. N_2 can also be transformed to plant available forms through the fertilizer manufacturing process. The chemical triple bond that exists between the two N atoms in N_2 is very strong. All the above processes that convert N_2 to plant useable forms, therefore, are highly energy intensive.

Nitrogen is an extremely reactive element with many of its possible transformations being depicted in the nitrogen cycle. The nitrogen cycle is the dynamic system in which nitrogen is transformed, or cycled, from one form to another. Total nitrogen in the soil-plant-water-atmosphere continuum is conserved, but amounts existing in various "pools", or forms, of nitrogen change with time and environmental conditions.

The vast majority of soil nitrogen (approximately 95 per cent) is found in soil organic matter. This organically-combined nitrogen is not immediately plant available, but can be converted to inorganic, available forms through the actions of soil microorganisms. This process, as previously discussed, is termed nitrogen mineralisation. Prior to the production of inorganic nitrogen fertilizers, most nitrogen for plant growth was supplied through leguminous N_2 fixation and nitrogen mineralisation of added animal manures and indigenous soil organic nitrogen. Sole reliance on nitrogen mineralisation from soil organic matter normally is not sufficient for crop production and may result in soil deterioration associated with the loss of organic matter. Nitrogen supplementation, whether from organic or inorganic fertilizer sources, is normally necessary for crop production.

Nitrogen, primarily in the form of nitrate, may be lost from soils through leaching. Nitrate, being an anion, is repelled by the negatively-charged cation exchange sites in soils. Since, the ion is not adsorbed and is highly water soluble, it will move downward in soils with percolating water. Nitrate leaching is most common in coarse, sandy soils receiving excess rainfall or irrigation.

Nitrate losses may also be increased by applying nitrogen fertilizers, whether inorganic or organic, in excess of a crop's requirement. Nitrate leaching should be prevented not only from economic, but also from environmental and health standpoints. Ingestion of waters high in nitrate has been implicated in gastrointestinal problems in adults and methemoglobinemia ("blue baby syndrome") in infants, though confirmed cases are rare. Near-surface aquifers normally are the most susceptible. Nitrate contamination of waters is usually localised and can be decreased through proper management.

Denitrification is the bacterial reduction of nitrate under anaerobic conditions to N_2 or N_2O gases. Under anaerobic conditions, most nitrate in soils may be denitrified in a period of a few days. Some scientists theorise that atmospheric N_2 is the result of denitrification over geologic time. Evidence also indicates that N_2O may be partially responsible for depletion of the protective ozone layer and is also a potent "greenhouse gas". Thus, loss of nitrate through denitrification not only results in an economic loss of plant available nitrogen, but may also have other detrimental effects. Combustion of fossil fuels also produces nitrous oxides that may contribute to this effect.

Nitrogen is an essential ingredient for the production of sufficient food for an expanding world population. Proper nitrogen management can decrease the potential for negative environmental impacts.

Phosphorus: Phosphorus in soil organic matter accounts for about 20 to 65 per cent of the total phosphorus found in soils. Therefore, phosphorus mineralisation from soil organic matter is an important source of available phosphorus for plant growth. Phosphorous ranks second to nitrogen as a limiting nutrient for plant growth. Although plant available forms of this element are anionic, phosphorus is immobile in soils with appreciable colloid content because it tends to be tightly bound to these tiny particles. Phosphorus may also form water insoluble compounds such as insoluble calcium phosphates in alkaline soils and insoluble iron and aluminum phosphates in acid soils.

The concentration of phosphorus in soil solution is normally much less than one part per million (ppm), even in fertilized soils, and often is only hundredths of a ppm in unfertilized soils.

Phosphorus fertilizers are normally produced through acidification of the mineral, apatite, found in high concentrations in some sedimentary deposits. Organic phosphorus sources, such as manure, may also be used. Manures, however, usually contain relatively large quantities of phosphorus relative to nitrogen. Care must be taken with manure additions so that excess phosphorus doesn't result in deficiencies of other nutrients, such as zinc, or contribute to soluble phosphorus in run-off waters.

Soluble phosphorus can be lost in surface run-off waters, but is usually found adsorbed to soil particles transported by erosion. Phosphorus in run-off has been implicated in eutrophication (excessive algal growth) of lakes and streams.

Potassium: Potassium is required by plants in amounts second only to nitrogen. Unlike nitrogen and phosphorus, potassium is not organically combined in soil organic matter. Different potassium-containing minerals, such as micas and feldspars, therefore, are the principal sources of potassium in soils. Clay-sized micas weather more rapidly to release potassium than feldspars because of their much greater surface area. Soils that contain considerable micaceous clays may be able to supply all of a crop's potassium requirement without fertilization. Acid, weathered soils are those most likely to be deficient in available potassium.

Calcium and Magnesium: Calcium is the predominant exchangeable cation in soils, even in the majority of acid soils, followed by magnesium. This occurs because of the large number of minerals in soils that contain calcium and/or magnesium. Actual plant deficiencies of these elements are infrequent because problems associated with soil acidity, such as aluminum toxicity, become limiting first.

Sulphur: Approximately 85 per cent of total soil sulphur is found in soil organic matter. Microbial mineralisation of the soil organic fraction is an important source of available sulphur for plant growth. Reactions of sulphur in soils are very similar to those of nitrogen. Sulphides, such as pyrite and other reduced forms of sulphur, are commonly unearthed in metal and coal mining. Upon exposure to oxygen, sulphuric acid which is produced through chemical and biological oxidation can result in soil acidification and acid mine drainage.

Soils may also receive sulphur through atmospheric deposition. Soils near large metropolitan areas may receive greater than 150 lbs S/acre/year from the combustion of fossil fuels. Volcanic eruptions can also emit large quantities of sulphur gases. Soils most commonly deficient in available sulphur are sandy, leached soils that are low in organic matter.

Micronutrients: Iron, zinc, manganese, copper, chlorine, boron, and molybdenum are classified as micronutrients. Micronutrients are plant essential elements that are required by plants in much smaller amounts than the other essential nutrients. Generally, less than 1 lb/acre of each micronutrient will be present in the aboveground portion of crops. This small quantity contrasts with the 200 lb/acre or more of nitrogen.

The total quantity of many micronutrients in soils doesn't necessarily relate to plant availability. Most soils, for example, will contain from 20,000 to 200,000 lbs total iron/acre to a depth of six inches, but may not be able to supply a crop with sufficient available iron for uptake of 1 lb/acre. Iron deficiency is usually associated with highly alkaline soils because iron solubility roughly decreases 1000-fold for each one unit increase in soil pH.

Zinc, manganese, and copper availabilities are also decreased by alkalinity and by high organic matter concentrations (>10 per cent). These three elements form very stable bonds with soil organic matter which decrease their availability.

Plant micronutrient deficiencies are becoming more widespread because of greater quantities required by higher yields and decreasing micronutrient impurities in fertilizers.

NUTRIENT DEFICIENCY AND TOXICITY SYMPTOMS IN TREE FRUIT

Growers often look at their orchards and suspect that something is not quite correct. Listed below are general guidelines for diagnosing symptoms that result from deficiencies or toxicities of certain elements. A tree may be deficient in these elements not only because they are scarce but also because an excess of other elements has prevented a balanced uptake of essential elements. The guidelines are not precise, and drastic changes in fertilizer practices should not be made based on a visual assessment. If you suspect a nutrient deficiency or toxicity, a leaf analysis should be conducted to confirm the visual symptoms. Information that follows was adapted from Childers, Shear and Faust, and Stiles and Reid. Elements required by plants are broken down into two broad categories: macronutrients and micronutrients. Plants need both to grow, flower, and fruit naturally. The distinction between the two groups is the quantities required by the plant. Macronutrients are needed in a larger amount and are usually expressed as per cent dry weight. The micronutrients are required in smaller quantities and are usually expressed in parts per million.

Macronutrients

Nitrogen:

- *Deficiency:* Symptoms appear as reduced top growth with short spindly shoots that have pale green to yellow leaves. Generally speaking, symptoms first become evident in the older leaves at the base of shoots. In stone fruits deficient leaves appear reddish and may exhibit a "shothole" effect as the condition worsens. Fruits, especially stone fruits, tend to be smaller and to mature earlier.
- *Toxicity:* Symptoms appear as an excessive amount of shoot growth accompanied by dark green foliage and delayed leaf drop in the fall. As nitrogen increases above the optimum, fruit colour is reduced and maturity is delayed. Red cultivars are less red, and yellow cultivars tend to remain green. In apples and pears flavour and storage life are reduced. Besides direct toxic effects, other physiological problems, such as corking and bitter pit, can occur in apples and pears because of higher nitrogen levels that are not in balance with calcium levels.

Magnesium:

- *Deficiency:* In their severest stages, symptoms may look similar to the marginal scorching associated with potassium deficiency, although this is very rarely observed in Pennsylvania. More characteristic is a fading of the green colour at the terminals of older leaves, progressing interveinally (between veins) towards the base and midrib of the leaf and giving the typical "herringbone" appearance. In pears, dark purplish islands of tissue surrounded by chlorotic bands may develop in the interveinal areas. As the growing season progresses, symptoms develop on progressively younger leaves and the older leaves fall off.
- *Toxicity:* Symptoms of excessive magnesium levels are not specific but usually appear as a deficiency of either potassium or calcium.

Micronutrients: The following elements are classified as micronutrients because they are required by plants in smaller quantities. With the exception of manganese, toxicities are very difficult to diagnose visually. Deficiency symptoms tend to be characteristic of the lack of a particular element.

Iron:

- *Deficiency:* Iron deficiencies are very common in plants. Initial symptoms are a loss of green colour in the very young leaves. While the interveinal tissue becomes pale green, yellow, or even white, the veins remain dark green. New leaves may unfold completely devoid of colour, but the veins usually turn green later.

- *Toxicity:* Although rare in the field, an excess of iron usually produces symptoms similar to those of manganese deficiency.

Manganese:

- *Deficiency:* Symptoms begin as chlorosis between the main veins starting near the margin of the leaf and extending towards the midrib. Symptoms can often be confused with those of iron and magnesium deficiencies. But unlike magnesium deficiency, manganese deficiency symptoms seldom develop so far as to produce interveinal chlorosis, the chlorosis normally being confined to leaf margins. The other distinguishing characteristic is that manganese deficiencies appear on the youngest leaves first, and the finest leaf veins do not remain green as they do with iron deficiencies.
- *Toxicity:* "Measles" is a disorder of apples, especially Delicious and Jonathan. It is caused in part by an excess of manganese accompanied by low calcium levels.

Boron:

- *Deficiency:* In most fruit crops, boron deficiencies show up in the fruit before appearing in the leaves. Symptoms in apples and pears are similar: gnarled, misshapen fruit caused by depressions usually underlaid by hard corky tissue. This symptom is often confused with bitter pit or corking caused by calcium deficiency. Researchers disagree whether the two symptoms can be told apart visually. Boron deficiency might be distinguished from bitter pit by the presence of pitting from the peel to the core, whereas in bitter pit the pitting usually occurs only at the calyx end and only very close to the skin. In some instances of boron deficiency, the entire surface is covered with cracks that have callused over, producing a russeted appearance. In plums, the symptoms appear as brown sunken areas in the fruit flesh, ranging in size from small spots to almost the whole fruit. The fruit usually colours earlier than normal, and falls. Gum pockets may also form in the flesh. In peaches, the flesh adjacent to the pit develops brown, dry, corky areas, and some fruits may crack along the suture. The most typical vegetative symptom is the death of the terminal growing points, resulting in a "witch's broom" appearance.
- *Toxicity:* Symptoms in apples include dieback of twigs, greatly enlarged nodes on 1-and 2-year-old twigs, early fruit maturity, internal breakdown, and dropping of fruit. Foliar symptoms occur first on the older leaves and include a yellowing along the midrib and the large lateral veins. In peaches vegetative symptoms include necrotic lesions on leaves, crinkling of margins and tips of leaves, reduced flower bud formation and set, and pit splitting.

Copper:

- *Deficiency:* Younger leaves appear stunted or misshapen, narrow, and slightly elongated with wavy margins. There may be some terminal dieback. Copper and zinc deficiencies often occur together and are aggravated in soils that have a high pH.
- *Toxicity:* Symptoms are almost non-existent under orchard conditions, but when present they may resemble those of zinc deficiency.

Zinc:

- *Deficiency:* Symptoms have often been described as a "rosetting" of leaves or "little leaf." Newly developing leaves are smaller than normal. Reduced shoot elongation keeps them close together, resulting in the rosette appearance. In severe cases, older leaves may drop, resulting in a more pronounced rosetting. In the early spring, the observer might notice a delayed foliation of lateral leaves on last year's shoots. This symptom has been confused with that of winter injury, but the distinguishing characteristic is that winter injury will also produce browning of the cambium.
- *Toxicity*: Symptoms are rare and most likely are masked by secondary symptoms resembling those of other micronutrient toxicities.

FOLIAR APPLICATION OF NUTRIENTS

The most efficient way to apply nitrogen, phosphorus, potassium, and magnesium is by ground application. Foliar applications of these elements should be viewed as temporary or emergency solutions only. Boron, zinc, copper, and manganese can be added by either foliar or ground application.

The foliar method is usually preferred because very small amounts are applied per acre. Since, formulations differ by manufacturer, it is essential to check compatibility of the material when mixing. Foliar application of calcium to control cork spot and bitter pit is dealt with in Cork Spot and Bitter Pit Fruit Disorders.

FERTILIZER APPLICATION NITROGEN

Nitrogen requirements are best determined by tree growth and performance. Soil tests for nitrogen have not proven useful in determining tree needs. As mentioned previously, nitrogen levels in a foliar analyses should be higher for young non-bearing trees.

Leaf nitrogen levels also tend to be higher in samples from trees carrying heavy crops. Biennial bearing trees in their off year or trees with a light crop generally have lower nitrogen levels. This is probably related to the inverse relationship between shoot growth and fruiting. Nitrogen cycle and its availability is a complex situation. Nearly 80 per cent of our atmosphere is composed of nitrogen. Some can be precipitated out during thunderstorms and rain events. Additionally, much nitrogen is recycled when organic matter decomposes. It has been estimated that for every 1 per cent of organic matter in the soil as much as 20 to 45 pounds of nitrogen per acre per year can be made available for plants to absorb. In addition, much of the nitrogen in the leaves can be remobilised back into the tree to be stored in the structural portions of the tree. This stored nitrogen is what supplies the initial tree growth in the spring. The larger the tree, the more potential nitrogen the tree can store. Conversely, this means that the new smaller trees on a per–tree basis will store less nitrogen. In apples, we have a better understanding of nitrogen levels based on cultivar.

Cornell University researchers suggest the following nitrogen levels in mature trees based on cultivar: 1.8 to 2.2 per cent nitrogen: Soft cultivars such as Cortland, Gala, Golden Delicious, Jerseymac, Jonagold, Jonamac, Jonathan, Macoun, McIntosh, Mutsu, Paulared, Spartan, Tydeman's Red, and other early ripening cultivars. 2.2 to 2.4 per cent nitrogen: Delicious, Empire, Idared, Liberty, Melrose, Rhode Island Greening, Rome Beauty, Stayman, York Imperial, and other varieties.

Application Rates: As a general rule, young trees may require about 0.01 to 0.04 pound of actual nitrogen per year of age up to 0.3 pound actual nitrogen per tree at maturity. These general levels should be adjusted up or down, depending on pruning, size of crop, cultural practices, indices presented above, and leaf analyses. Young non-bearing trees should grow up to 24 inches annually. Mature and heavily bearing trees should grow less, with apple and pear trees growing 12 to 18 inches; apricot, cherry, plum, and prune trees 15 to 18 inches; and peach and nectarine trees 18 to 24 inches.

Form of Nitrogen: Fruit trees respond to any form of nitrogen fertilizer. Avoid using ammonium sulphate, urea, or ammonium nitrate if pH is below 6.0. Urea sprays may help increase fruit set but will not supply trees with all the nitrogen they require. Pear trees do not respond well to urea sprays. Applications of urea sprays to stone fruits are ineffective during the growing season. Calcium nitrate sprays should be avoided to discourage corking.

Time of Application: Traditionally, we have said that nitrogen should be applied no later than 4 to 6 weeks before bloom. Recent research in the Pacific Northwest, however, indicates that application at this time results in the nitrogen chiefly accumulating in the foliage of the trees with little going into the flowers and developing fruitlets to help fruit set. Studies have shown that the majority of the nitrogen used to set fruit comes from the reserves within the tree and spring–applied nitrogen does not reach the developing fruit in time to be effective. In studies in apples in Oregon, late–summer to early fall nitrogen application was found to be preferentially translocated to the roots and the flower buds where it was mobilised from the roots and available in the flower buds next spring to increase flower strength and fruit set capabilities.

The theory that nitrogen should not be applied after the middle of July may be incorrect; however, we have no sound research on this different application timing in Pennsylvania. Until we have adequate information, the safest method is to continue your past fertilization practices or to experiment only on a small scale.

Phosphorus: Phosphorus is important to plant growth because it is a catalysing agent that induces metabolic reactions. It permits the plant to use nitrogen and to develop seeds. Failure of seeds to set often results in abortion of the young fruit and in misshapen fruit. Preplant soil test recommendations are made to attempt to raise soil levels to 100 pounds available phosphorus per acre.

Application Rates: Rates should be based on either leaf analysis or soil analysis. Soil levels of less than 100 pounds phosphorus per acre indicate that an application of phosphorus is needed. Leaf analysis levels of less than 0.18 per cent in apples and pears; 0.15 per cent in peaches and nectarines; 0.23 per cent in cherries; and 0.09 per cent in plums indicate a need for phosphorus. Form of phosphorus: The effectiveness of special forms of phosphorus under Pennsylvania conditions has not been documented.

Time of Application: Application may be made anytime during the year in established orchards. Preferably, applications should be broadcast before the trees are planted and turned under with the previous crop. This aids in getting phosphorus down into the root zone. ***Potassium:*** Potassium is believed important for maintaining water turgor in leaves and for functioning in the opening and closing of stomates. It is available in the soil as a cation; at excessive levels it competes with calcium and magnesium for uptake by the plant.

Application Rates: If the potassium level in the foliage is over 1.5 per cent, a response to potassium application is doubtful regardless of the soil test, unless leaf nitrogen or soil magnesium is too high. When soil test values exceed 4.5 per cent base saturation of potassium, then no additional potassium is needed. Values lower than this call for an application. Form of potassium: Any added value of one type of potassium fertilizer over another has not been determined. Time of application: Same timing as for phosphorus.

Calcium: Calcium plays a vital role in reducing the incidence of corking and bitter pit. Low soil–test values are often found in very sandy or shaley soils, but such values usually do not indicate a need to apply calcium or lime. Calcium in the form of limestone is important for maintaining soil pH.

Improper soil pH can lead to deficiencies or toxicities of other nutrients. The need for lime is best determined by a soil test. Calcium, like potassium and magnesium, is available to the plant as a cation, and an excess of potassium or magnesium can reduce calcium uptake. Application rates: Soil application depends on soil pH, buffer pH, and depth of pH change desired.

Calcium chloride applied as a foliar spray is recommended to prevent corking and bitter pit in the fruit. Form of calcium: Growers should base their limestonepurchase decisions on the price per ton of calcium carbonate equivalent, including spreading costs. Most limestone contains some impurities, the most common being magnesium. Continual use of dolomitic or high–magnesium lime can cause problems in a tree's uptake of calcium.

Dolomitic lime should not be used unless the soil test or the leaf analysis indicates a need for magnesium. Time of application: In preplanting situations lime is most effective when broadcast and incorporated at least 6 to 12 months prior to planting. In established orchards any time is suitable, although postharvest or early spring applications allow rains to move lime into the soil. Regardless of how much rain accumulates, the process of raising the soil pH is slow, with the effects of lime moving downward at about 1 inch per year.

Magnesium: Leaf analysis values below 0.2 per cent in apples and plums, 0.3 per cent in peaches and pears, and 0.49 per cent in cherries suggest that trees may respond to applications of magnesium. In soil tests, when the ratio of the percentage base saturation of magnesium to potassium is less than 2.0, magnesium applications may be needed.

Application Rates: Foliar applications of magnesium sulphate at 10 pounds per acre should be viewed as quick but temporary measures. A more permanent solution is to apply a magnesium containing limestone at rates recommended in the soil analysis. Form of magnesium: Magnesium is normally found in all but the purest limestone. The Penn State soil test results give a separate recommendation for magnesium. Growers should examine the purity of the limestone to determine whether the percentage of magnesium contained in it will also satisfy the magnesium requirement.

Time of Application: Magnesium sprays are effective only when applied during the growing season. Applications may be single or split (*i.e.*, applied at petal fall or in the first two cover sprays). Foliar sprays appear to be most effective when applied separately from pesticide applications. Ground applications can be made when lime is applied.

Boron: Boron is a nutrient often found to be low or deficient in orchards. Deficiency symptoms are observed more often in apple, plum, and pear trees than in peach and cherry trees. In apple trees, a deficiency may be expressed as internal cork in the fruit, as a dieback of shoots, and as bark necrosis. In pear trees, a deficiency may be expressed as internal cork, withering of blossoms, or poor fruit set. In peach trees, toxicity symptoms appear as necrotic lesions on leaves, crinkling of margins and tips of leaves, reduced flower bud formation and set, and pit splitting.

Tissue test values less than 35 ppm in apples indicate a shortage of boron. Values between 35 and 60 ppm indicate sufficiency. In the soil, values less than 0.5 ppm are low, while values between 0.5 and 1.0 ppm indicate a sufficient level. Caution: Excessive boron can be extremely toxic. Leaf values of over 80 ppm or soil values of over 1.0 ppm are excessive. *Application Rates:* In orchards where boron is low, apply 0.8 to 1.6 pounds per acre of actual boron (4–8 lb/A of Solubor 20.5 per cent B) in two separate sprays at bloom, petal fall, or first cover; or apply a single postharvest foliar spray of 1.6 pounds per acre of actual boron. For trees more than 3 years old, apply boron annually to the soil.

Apply 0.12 pound of actual boron per acre for 4-year-old trees. For each additional year of tree age up to 16 years, increase the rate by 0.02 pound per acre of actual boron. Use the lower rate as a maintenance programme when no leaf analysis has been made.

Eight pounds of Solubor per acre is recommended for proven cases of low boron. Form of boron: Boron is most effective when applied as a foliar spray. Special formulations such as Boro–spray or Solubor should be used. Agricultural borax is adequate for ground applications. Time of application: Soil applications can be made anytime. Foliar applications can be made during bloom or postharvest while leaves are still green and active on the tree.

Copper: Copper is a catalysing element in plant metabolic reactions. Foliar analysis is best for determining copper deficiency. Values below 5 ppm indicate the need to apply copper. Application rates: When leaf analysis indicates a deficiency, 4 to 6 pounds per acre of copper sulphate is recommended. Form of copper: Copper sulphate (22 per cent Cu) is the most readily available and cheapest material to use. Time of application: Foliar applications should be made during the dormant season or after the fruit is off and while the leaves are still active and green. Soil applications can be made anytime. ***Zinc:*** Deficiencies of zinc are becoming more common in Pennsylvania. Small leaves clustered on the end of shoots indicate zinc deficiency but may be confused with winter injury. Zinc applications are recommended when leaf levels are below 20 ppm. Foliar applications of zinc sulphate should only be made during dormancy or postharvest. Do not apply zinc sulphate with oil or apply within 30 days of the application of oil.

Bibliography

A.K. Mandal and G.L. Gibson : *Forest Genetics and Tree Breeding*, CBS Publicatoin, Delhi, 2002.

Benu Singh : *A Modern Book on Forestry and Horticulture*, Vista International Publication, Delhi, 2010.

Benu Singh : *A Survey of the Forestry Research*, Vista International Publication, Delhi, 2011.

P R Gilbert : *Agricultural Biotechnology, Forestry and Products*, Anmol Publication, Delhi, 2008.

Praveen Taank : *Advances in Forestry Research in India*, Cyber Tech Publication, Delhi, 2008.

R.K Bhardwaj : *Advances in Forestry Research in India*, Oxford Book Company, Delhi, 2008.

R.K. Singh : *A Manual of Forest Act Conservation*, International Publication, Delhi, 2000.

Ram Parkash : *Advances in Forestry Research in India*, International Book Distributors, Delhi, 2002.

Richard P Tucker : *A Forest History Of India*, Sage Publication, Delhi, 2011.

Roger Jeffery and Nandini Sundar : *A New Moral Economy for India's Forests?*, Sage Publication, Delhi, 1998.

S P Singh : *Advances in Horticulture and Forestry*, Scientific Publication, Delhi, 2006.

S S Negi : *Agro Forestry Handbook*, International Book Dist, 2007.

S. John Joseph and B.S. Nagarajan : *A Paradigm Shift in Forest Management*, International Book Dist, 2004.

S.K. Mukherjee : *Advances in Forestry Research in India*, International Book, Delhi, 2004.

S.P. Singh : *Advances in Horticulture and Forestry*, Scientific Publication, Delhi, 2001.

Sandeep Kumar and Matthias Fladung : *Molecular Genetics and Breeding of Forest Trees*, International Publication, Delhi, 2005.

Solon L. Barraclough and Krishna B Ghimire : *Agricultural Expansion and Tropical Deforestation*, Viva Books, Delhi, 2006.

T.V. Sathe : *A Textbook of Forest Entomology*, Daya Publication, Delhi, 2008.

Ajay Sharma: *A Text Book of Forest Trees Entomology*, International Book Distributers, Delhi, 2007.

B Venkatesh; B K Purandara and K S Ramasastri: *Forest Hydrology*, Capital Publications, Delhi, 2007.

Benu Singh: *A Modern Book on Forestry and Horticulture*, Vista International Publications, Delhi, 2010.

Benu Singh: *A Survey of the Forestry Research*, Vista International Publications, Delhi, 2011.

Charlotte Streck, Robert O'Sullivan, Toby Janson-Smith and Richard Tarasofsky: *Climate Change and Forests: Emerging Policy and Market Opportunities*, Concept Publications, Delhi, 2009.

Falguni Chakrabarty: *Adaptation of the Santals to the Hill-Forest Environment,* APH Publications, Delhi, 2012.

G P D Vyas: *Community Forestry*, Agrobios Publications, Jodhpur, 2006.

Hemant Kumar Gupta: *Joint Forest Management: Policy, Participation and Practice in India*, International Book Distributers, Delhi, 2006.

J B Lal: *Forest Management: Classical Approach and Current Imperatives*, Natraj Publications, Delhi, 2007.

J.P. Pascal & B.R. Ramesh: *A Field Key to the Trees and Lianas of the Evergreen Forests of the Western Ghats (India)*, Institute Francais De Pondichery, 1997.

John W Bruce: *Community Forestry: Rapid Appraisal of Tree and Land Tenure*, Daya Publications, Delhi, 2005.

K. Jayaraman: *A Handbook on Statistical Analysis in Forestry Research*, Kerala Forest Research Institute, 2001.

K.D. Singh, Bhaskar Sinha and Pravat C. Sutar: *Community Based Forest Management With Micro Enterprise For Poverty Reduction*, Bishen Singh Mahendra Pal Singh, Dehradun, 2009.

L K Jha and P K Sen Sarma: *A Manual of Forestry Extension Education*, A.P.H Publications, Delhi, 2008.

M P Singh; J K Singh; Reena Mohanka and R B Sah: *Forest Environment and Biodiversity*, Daya Publications, Delhi, 2007.

M R K Jerram: *Text Book on Forest Management*, International Book Distributers, Delhi, 2006.

M.M. Deka: *Joint Forest Management in Assam*, Daya Publications,

Delhi, 2002.

Index

I

L

M

N

O

P

R

S

T

U

V

W